ENERGY EFFICIENCY IN BUILDINGS: PROGRESS AND PROMISE

American Council for an Energy-Efficient Economy
Series on Energy Conservation and Energy Policy

Series Editor, Carl Blumstein

Financing Energy Conservation
Energy Efficiency in Buildings: Progress and Promise

ENERGY EFFICIENCY IN BUILDINGS: PROGRESS AND PROMISE

Eric Hirst
Jeanne Clinton
Howard Geller
Walter Kroner

F. M. O'Hara, Jr., Editor

American Council for an Energy-Efficient Economy
1001 Connecticut Ave., N.W.
Washington, D.C. 20036

1986

Library of Congress Cataloging-in-Publication Data

Energy efficiency in buildings.

Bibliography: p.
Includes index.
1. Buildings—Energy conservation.   I. Hirst, Eric. II. O'Hara, Frederick M.
III. American Council for an Energy-Efficient Economy.
TJ163.5.B84E5425   1986   696   86-7998
ISBN 0-918249-04-X

Library of Congress Catalog Card Number: 86-7998
ISBN: 0-918249-04-X

First published in 1986.

American Council for an Energy-Efficient Economy
1001 Connecticut Ave., N.W., Washington, D.C. 20036

Printed in the United States of America.

10 9 8 7 6 5 4 3 2 1

Cover art by Allison Turner

CONTENTS

Part III. RESEARCH AND PROGRAM DEVELOPMENT:
AGENDAS, THEMES, AND VISIONS

WHY ANOTHER BOOK ON ENERGY EFFICIENCY?

Does it make sense to publish a book on energy efficiency in 1986? Oil supplies are abundant, oil prices are low, and OPEC is in disarray. Many gas and electric utilities have excess capacity and are competing vigorously with each other for new customers. To some, this may seem an unlikely time to devote more attention to energy efficiency in residential and commercial buildings.

But how likely is it that the current situation will persist? Although fuel prices are currently stable and supplies are ample, many predict that this will change within a few years because of the physical limitations on nonrenewable energy resources and the likely growth of the worldwide economy. A few statistics suggest future changes: (1) U.S. oil reserves are falling and can provide only nine years of supply at current production levels. (2) The U.S. imported 30% of its oil supplies in 1984 at a cost of $50 billion. (3) Almost two-thirds of world oil reserves are in the volatile Middle East, a fraction that is rising as nonOPEC nations rapidly consume their indigenous oil supplies.

During the past decade, energy efficiency improvements were the most important energy resource in the U.S., exceeding the contributions from coal, nuclear, oil, and natural gas. To a large extent, energy conservation helped create the present oil glut and stable oil prices.

Further increases in energy efficiency will stretch supplies of nonrenewable resources, cushion our economy against future fuel shortages or price hikes, and provide more time to develop alternative fuels. In addition to these long-term global objectives, there are strong pocketbook reasons to increase energy efficiency levels. Improved energy efficiency can drastically reduce the operating costs of residential and commercial buildings; can reduce the investment requirements of electric utilities, thereby making capital available to strengthen the U.S. economy; can increase local and regional employment; can reduce the adverse environmental impacts of fuels extraction, transportation, and conversion; can reduce U.S. dependence on foreign fuel supplies; and can improve our nation's balance of payments.

We should use the next few years as a window of opportunity, a time to explore and develop technologies and programs that will improve energy efficiency in cost-effective ways. We believe that it is extremely important to understand what has been accomplished in energy conservation, identify gaps in our knowledge and deficiencies in program implementation, and get on with the job of upgrading our residential and commercial buildings to efficiency levels justified by today's fuel prices. Instead of becoming complacent as we watch OPEC falter, we should develop new technologies, policies, and programs to stimulate greater energy efficiency.

PURPOSE AND ORGANIZATION OF BOOK

The purpose of this book is twofold: to review current knowledge on energy use and efficiency in residential and commercial buildings and to suggest important research and program topics for future study.

The introductory chapters set the stage for subsequent discussions of what we now know and where we want to go. We first review overall patterns of energy use in residential and commercial buildings and the dramatic changes in energy trends after 1973, including the roles of government, utility, and private sector efforts in making these changes. We next discuss reasons for continuing research and programs to improve energy efficiency in buildings and note the complexity and diversity among buildings in their design, construction, operation, maintenance, and use. We then turn our attention to how much has been accomplished and learned about reducing energy use in buildings since the 1973 oil embargo. Finally, we offer many suggestions that merit attention in both the short and long terms. The short-term proposals flow logically from current research and programs on energy efficient buildings. The long-term agenda (including several visions of the future) covers ideas that, in some ways, require changes in how we view our built environment and the social institutions under which buildings are constructed and operated.

AUDIENCE FOR BOOK

In planning, organizing, writing, and editing this book, we struggled with a basic question: Who is our audience? Our goal is to reach two groups. The first includes those responsible for R&D funding and conservation program decisions (i.e., policy makers and program managers). We hope that these officials in government agencies, electric and gas utilities, private companies, and architecture and engineering firms will find the book helpful in defining future program priorities and funding levels. The second group includes energy professionals and others interested in the field of building energy conservation. We hope that our fellow researchers, teachers, and program staff will appreciate and learn from the reviews of past accomplishments and will accept some of the specific suggestions developed here. We also hope that readers will be stimulated by the book to develop their own ideas for improving energy efficiency in buildings.

This is not a book that one reads from beginning to end. Researchers and conservation decisionmakers will probably read those chapters that most interest them. Thus, the book should be viewed primarily as a reference (like an encyclopedia), rather than as a text (in which each chapter logically follows its predecessors).

AUTHORS

We are not the first group to suggest additional research in these areas. The Electric Power Research Institute, Gas Research Institute, American Institute of Architects, U.S. Department of Energy, and U.S. House of Representatives Committee on Science and Technology have all convened meetings and/or published research agendas. This book differs from prior efforts in one important respect: ours is a personal effort, the result of 15 months of work by four individuals, while the others generally represent institutional statements.

The four of us represent a diversity of backgrounds, interests, and occupations. Jeanne Clinton is a planner and manager, with experience in a state energy office, municipal electric utility, and now the energy consulting firm of Barakat Associates. Howard Geller is a mechanical engineer with international research experience, currently senior researcher and Associate Director of the American Council for an Energy-Efficient Economy. Eric Hirst is also a mechanical engineer, with energy analysis experience at the Oak Ridge National Laboratory, a state energy office, and the Federal Energy Administration. Walter Kroner is a practicing architect involved with energy-conscious building design, architecture professor, and director of the Center for Architectural Research at Rennselaer Polytechnic Institute. We believe that the variety of perspectives we bring to this book make it interesting, lively, and informative to our readers.

Our personal interests and perspectives, plus the inevitable time constraints, led to the omission of some topics. For example, there is little within this book on active solar systems, on thermal storage, or on utility load management technologies.

AMERICAN COUNCIL FOR AN ENERGY-EFFICIENT ECONOMY

The American Council for an Energy-Efficient Economy, publisher and sponsor of this book, is a nonprofit research and educational organization located in Washington, D.C. The ACEEE sponsors research, publishes energy-related consumer guides, and participates in public discussions of energy policy and research. Perhaps ACEEE's best known activity is sponsorship of the biennial conference on *Energy Efficiency In Buildings*, held at the Santa Cruz campus of the University of California. To some extent, this book is an outgrowth of the 1984 conference, *Doing Better: Setting an Agenda for the Second Decade*, in that it draws upon the conference papers and on the attendees themselves.

ACKNOWLEDGMENTS

Writing a book, as we learned the hard way, is an enormous undertaking that takes a great deal of time, effort, and especially determination. Happily, preparation of this book involved interaction with many wonderful people without whose help the book would never have been completed. Carl Blumstein first thought of the book in early 1984 while organizing the 1984 ACEEE Santa Cruz Summer Study. It was Carl who developed the initial ideas for the book's major themes. He and Linda Schuck helped identify and assemble the four ACEEE Fellows who wrote the book.

Many people reviewed portions of the book in draft form. We thank Jon Veigel, Maxine Savitz, John Millhone, Debbie Bleviss, Alan Meier, Carl Blumstein, Paul Stern, Rob Socolow, Gautum Dutt, John Morrill, David Moulton, John Armstrong, Arthur Rosenfeld, William Fulkerson, and Marc Ledbetter for their helpful comments. We are especially grateful to Roger Carlsmith, Jeff Harris, Tom Potter, and Grant Thompson for their reviews of the entire draft book.

We are indebted to Fred O'Hara, who edited the book. This involved careful reviews and revisions to reconcile the individual views and writing styles of the four authors. Debbie Barker and Marjie Hubbard typed draft after draft of the book. Linda O'Hara did the final copy editing, Michael O'Hara performed the makeup, and Jamie Crigger prepared the final typeset version of the text. Andy Loebl provided moral and administrative support at ORNL throughout preparation of the book.

Finally, we are grateful to the many organizations that supported the 1984 summer study and made this book possible: Bonneville Power

Administration, Electric Power Research Institute, Gas Research Institute, Lawrence Berkeley Laboratory, Michigan State University, Oak Ridge National Laboratory, Pacific Gas and Electric Company, Southern California Edison Company, Standard Oil Company (Ohio), U.S. Department of Energy, University of California, Western Area Power Administration (all sponsors); California Energy Commission and North Carolina Alternative Energy Corporation (both major contributors); and Owens-Corning Fiberglas Corporation, Governor's Energy Council of Pennsylvania, and Time Energy Systems, Inc. (all contributors).

LIST OF ACRONYMS

ACEEE	American Council for an Energy-Efficient Economy
AIA	American Institute of Architects
ASHRAE	American Society of Heating, Refrigerating and Air-Conditioning Engineers
BECA	Buildings Energy Compilation and Analysis
BEPS	Buildings Energy Performance Standards
BPA	Bonneville Power Administration
BTECC	Building Thermal Envelope Coordinating Council
CAC	central air conditioner
DOE	U.S. Department of Energy
DSM	demand side management
EER	energy efficiency ratio
EES	Energy Extension Service
EIA	Energy Information Administration
EMS	energy management system
EPRI	Electric Power Research Institute
ESCO	energy service company
GAO	U.S. General Accounting Office
GNP	Gross National Product
GRI	Gas Research Institute
HERS	home energy rating system
HPWH	heat pump water heater
HUD	U.S. Department of Housing and Urban Development
HVAC	heating, ventilating, and air conditioning system
IAQ	indoor air quality
ICP	Institutional Conservation Program
IES	Illuminating Engineering Society of North America

k	thousand
LBL	Lawrence Berkeley Laboratory
LIHEAP	Low-Income Home Energy Assistance Program
M	million
NAHB	National Association of Home Builders
NCAEC	North Carolina Alternative Energy Corporation
NPPC	Northwest Power Planning Council
NIBS	National Institute of Building Sciences
NBECS	Nonresidential Buildings Energy Consumption Survey
OPEC	Organization of Petroleum Exporting Countries
ORNL	Oak Ridge National Laboratory
OTA	Office of Technology Assessment
PSC	public service commission
PUC	public utilities commission
PVEA	Petroleum Violation Escrow Account
Q	quadrillion (e.g., QBtu = 10^{15} Btu)
R	thermal resistance of a building or building component
R&D	research and development
RCS	Residential Conservation Service
RECS	Residential Energy Consumption Survey
SECP	State Energy Conservation Program
SEO	state energy office
SERI	Solar Energy Research Institute
TVA	Tennessee Valley Authority
U	inverse of R (1/R), building thermal conductance
VAV	variable air volume
WAP	Weatherization Assistance Program

I

BACKGROUND

1

PROGRESS IN ENERGY CONSERVATION
DURING THE PAST DECADE

Energy use in residential and commercial buildings is a complicated and fascinating subject, primarily because so many factors affect energy use and efficiency (Fig. 1.1). The three chapters in Part I provide background for subsequent discussions of current knowledge about, and future direction for, building energy efficiency. This chapter reviews recent trends in building energy uses and the roles of governments, utilities, and private firms in achieving the substantial efficiency gains that occurred since the early 1970s. Chapter 2 discusses reasons why, in spite of the impressive past progress, additional efforts to further improve energy efficiency are warranted. Chapter 3 explains the context in which buildings are designed, constructed, operated, and used.

A. PATTERNS OF ENERGY USE IN BUILDINGS

Energy use in residential and commercial buildings totaled 26 QBtu on a primary basis in 1983, equal to 36% of total U.S. energy consumption (Energy Information Administration 1984a)*. Residential buildings accounted for 15 QBtu (21% of national energy use), while commercial buildings accounted for 11 QBtu (15% of national energy use). Space heating is the dominant end-use in both residential and commercial buildings, accounting for more than 40% of the total in each (Fig. 1.2). Air conditioning and lighting are the next most important uses in commercial buildings, accounting for 20–25% each. These uses are minor in residential buildings, where water heating is the second most important use, accounting for almost 20% of the total.

---

*Primary energy units include losses associated with production of electricity (roughly 11,500 Btu/kWh). End-use units, rarely used here, do not (3412 Btu/kWh).

Although the two building sectors differ substantially in their energy use by end-use, they are quite similar in terms of fuel mix (Fig. 1.3). Electricity (measured on a primary energy basis) is the dominant fuel type, accounting for more than half the total in both sectors.

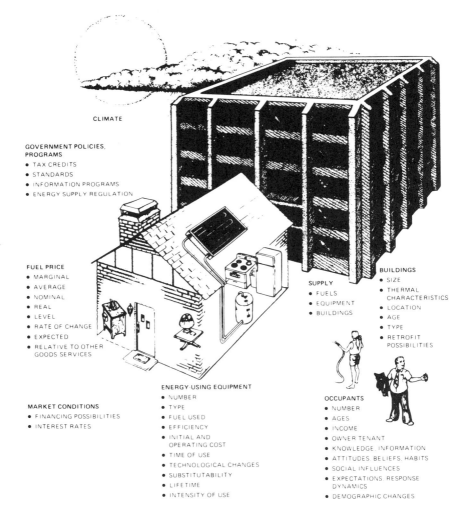

Fig. 1.1. Many factors affect energy use in residential and commercial buildings. These include characteristics of available fuels, building shell and equipment, government policies and programs, occupants, owners, and managers.

Commercial sector energy use is dominated by a few major building types; retail/wholesale, education, and office buildings account for almost two-thirds of the sector total (Fig. 1.4). In the residential sector, single-family homes are the major energy users (Fig. 1.4).

These summary statistics indicate the nature of energy use by end-use, fuel, and building type within the residential and commercial sectors. Another important variable among commercial buildings is their total energy use per unit floor area. For example, a survey of commercial buildings conducted in 1979 shows that health care facilities typically consumed 262 kBtu/ft^2, office buildings 124 kBtu/ft^2, and retail buildings 87 kBtu/ft^2 (Energy Information Administration 1984b; Pacific Northwest Laboratory 1984). In seemingly identical residences, energy use can vary by a factor of two or more depending on occupant behavior (Chapter 8). Chapters 4 through 8 examine in greater detail the technical and behavioral factors affecting energy use in buildings.

B. HISTORICAL TRENDS IN BUILDING ENERGY USE

In addition to the considerable variation across buildings, dramatic shifts have occurred in overall building energy use during the past few decades. Between 1950 and 1973, building energy use grew at a steady rate of about 4.4%/year. Between 1973 and 1984, energy use experienced

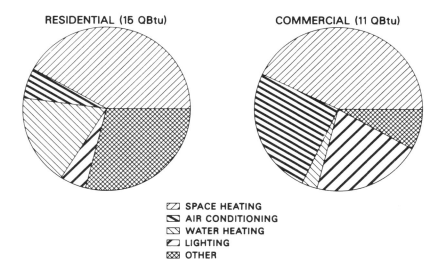

RESIDENTIAL (15 QBtu)   COMMERCIAL (11 QBtu)

- ▨ SPACE HEATING
- ◧ AIR CONDITIONING
- ◩ WATER HEATING
- ▱ LIGHTING
- ▨ OTHER

Fig. 1.2. Energy use in residential and commercial buildings by end use, 1980.

periods of increase as well as decline, such that primary energy use in 1983 was only 8% higher than it was in 1973 (equivalent to an overall growth rate during this post-oil-embargo period of 0.7%/year) (Energy Information Administration 1985).

Of course, the level of residential and commercial activities have increased substantially in the U.S. during the past 20 years. Consequently, as Figs. 1.5 and 1.6 show, both residential energy use per household and commercial energy use per unit of floor area declined during the past decade after increasing throughout the 1960s.

Table 1.1 shows residential end-use energy consumption during the past 20 years with normalizations for both the number of households and climate. While overall residential energy use declined 13% from 1972 to 1982, consumption per household dropped 31%. However, part of this drop was caused by a shift in population to the South and West, where energy use per household is below the national average because the climate is milder. After correcting for climatic differences, residential energy intensity (in terms of Btu per household per degree day) dropped 27% from 1972 to 1982. These reductions in residential energy consumption during the past decade were in sharp contrast to trends during the 1950s and 1960s. Quantitatively, residential end-use energy consumption in 1982

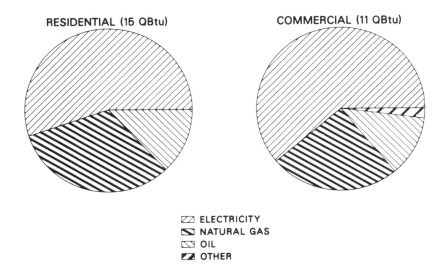

Fig. 1.3. Energy use in residential and commercial buildings by fuel, 1980.

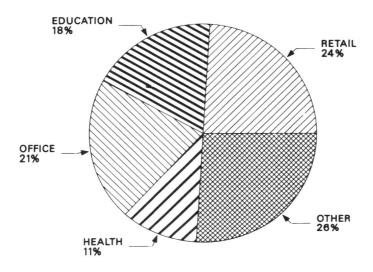

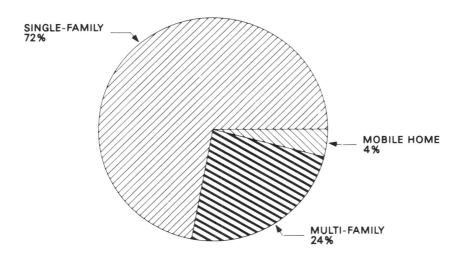

Fig. 1.4. **Energy use by building type in the commercial (top) and residential (bottom) sectors, 1980.**

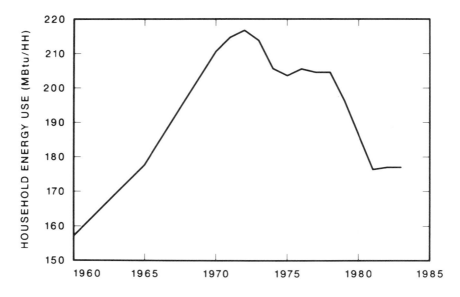

Fig. 1.5. Residential energy use per household, 1960–1983.

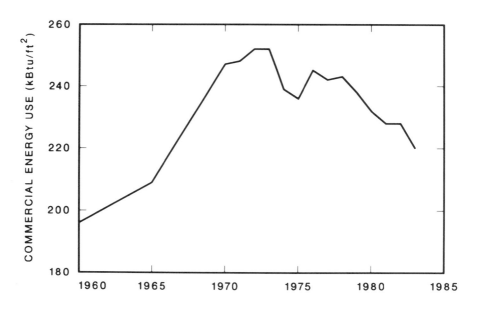

Fig. 1.6. Commercial energy use per unit floorspace, 1960–1983.

Table 1.1. Residential end-use energy consumption in the U.S.[a]

Year	Consumption (trillion Btu)	Consumption per household (MBtu/HH)	Consumption per household per heating degree day (kBtu)
1962	7,190 (100)	131.3 (100)	26.4 (100)
1972	10,108 (141)	151.6 (115)	30.8 (117)
1982	8,761 (122)	104.9 (80)	22.4 (85)

[a]Numbers in parentheses are an index with 1962 = 100.
Source: Energy Information Administration (1984b).

would have been about 60% greater had historical trends in household formation and household energy intensity continued on their 1962—72 paths.

Adams et al. (1984) analyzed changes in residential energy use from 1972 to 1982 to explain the determinants and manifestations of these changes. Their analysis suggests that about 25% of the total efficiency gain was produced by improvements in the shell (structure) of both new and existing homes, 20% was produced by decreases in household size, 10% was produced by more efficient appliances, 15% was produced by increased wood use, and 30% was produced by changes in household energy use practices (such as thermostat settings) and other factors.

The Department of Energy's Policy Office (1985) analyzed trends in U.S. energy use between 1972 and 1984. They concluded that 1984 residential energy use (15 QBtu) was 10 QBtu below what the pre-1972 trends would suggest. All of this decline was caused by greater efficiency per household. Similarly, 1984 commercial energy use (11 QBtu) was 7.2 QBtu below what historical trends would suggest. About 20% of this reduction was caused by slower growth in commercial floor area, while the remaining 80% was caused by reductions in energy use per unit floor area.

Between 1960 and 1972, primary energy use per unit of floorspace in the commercial sector increased 25%. During the subsequent decade, this measure of energy intensity declined by 11%. On an end-use basis, commercial energy use per unit of floor area dropped 22% from 1972 to 1982 (Energy Information Administration 1984a). Overall, primary energy use in the commercial sector in 1982 would have been at least 50% greater than it actually was had historical trends from 1960 to 1972 in floor-space growth and energy intensity continued.

The preceeding discussion shows that building energy efficiency improved during the past decade but does not explain why. Overall, our understanding of these recent changes and their determinants (i.e., the relative importance of fuel price increases, changes in economic growth and composition, technological advances, and other noneconomic forces, such as government regulations and utility conservation programs) is limited.

Analysis of total U.S. energy use (including industrial and transportation uses as well as building uses) conducted at the Oak Ridge National Laboratory (ORNL)(Hirst et al. 1983) suggested that energy use in 1982 (71 QBtu) was almost 35 QBtu lower than it would have been had historical (pre-embargo) trends continued unchanged. Nearly half the reduction was caused by slower growth in the GNP after 1973. The remaining 19 QBtu reduction was caused by improved energy efficiency, with roughly 40% caused by higher energy prices and the rest (13% or 4.5 QBtu) caused by the effects of government and utility conservation programs and nonenergy price factors (accelerated trends away from energy intensive industries, non-price-induced technological advances, energy shortages, deregulation of energy using industries, and increasing public awareness of the energy "crisis").

The results suggest that the sharp changes in energy use during the past decade were produced primarily by changes in economic activity and rising fuel prices and secondarily by other factors. Although the details differ from sector to sector, the qualitative effects surely apply to the residential and commercial sectors as well as to the economy as a whole. Recent trends in the residential sector appear to have been caused more by fuel price increases and "other" factors and less by economic growth than for the economy as a whole. Energy use changes in the commercial sector were more nearly parallel to those for the economy as a whole.

C. GOVERNMENT AND UTILITY
CONSERVATION PROGRAMS

Since the 1973 embargo, agencies at all levels of government as well as gas and electric utilities instituted a variety of programs aimed at improving energy efficiency of new and existing buildings.

At the federal level, annual expenditures on energy conservation research and programs by the U.S. Department of Energy (DOE) and predecessor agencies increased from zero in 1970 to $10 million (in 1972 dollars) in 1973, $490 million in 1980, and then declined to about $200 million after 1981. In addition to supporting technical research on energy use in building structures and equipment, DOE developed information, financial-incentive, and regulatory programs to stimulate energy conservation in buildings (Chapter 9).

For example, the Federal Energy Administration developed and tested a computerized home energy audit program in 1974, to provide site specific advice to households concerning the potential energy savings for certain retrofit measures. In 1978, Congress passed legislation requiring major gas and electric utilities to offer their residential customers onsite home energy audits. During the first three years of operation, this Residential Conservation Service audited about three million homes. Beginning in 1984, some utilities began offering commercial and apartment building audits in response to subsequent federal legislation.

In another example, a 1976 federal law requires standardized efficiency testing for several new residential products, including refrigerators, air conditioners, and water heaters. These products must display energy-efficiency and -cost labels in stores. Recently, the federal government has backed away from actively promoting energy conservation particularly through regulatory programs, such as building efficiency standards or appliance efficiency standards.

The federal government and some state governments provide tax credits for residential conservation expenditures. From 1978 through 1983, the conservation credit was claimed on an average of 4 million federal tax returns per year. The federal tax credit is worth 15% of conservation expenditures up to $2000 spent. From 1978 to 1983 the Federal government provided individuals with tax credits valued at $2.3 billion.

Utilities, especially during the past few years, have begun a variety of programs to encourage conservation. Virtually all utilities distribute promotional and educational materials and sponsor audit programs. Many utilities offer financial incentives to customers who purchase and install energy efficient equipment. For example, rebates are offered to commercial customers for installing high efficiency lighting and HVAC systems and to residential customers for purchase of high efficiency air conditioners and heat pumps. Financial incentives from utilities for conservation and peak-load reduction are available to 60% of U.S. households (Dickey et al. 1984).

Some utilities have started innovative programs to stimulate greater energy conservation among their customers. For example, General Public Utilities (New Jersey and Pennsylvania) hires contractors to retrofit homes in selected areas at no cost to the homeowner. The contractor is paid by the utility on the basis of the actual electricity savings achieved. Utility conservation programs as well as utility attitudes towards conservation are examined in Chapter 10.

D. PRIVATE SECTOR CONSERVATION EFFORTS

The progress made in improving energy efficiency in buildings is partly a result of impressive technological advances in building envelopes and the equipment used within buildings; see Chapters 4 to 7.

Many private companies manufacture, sell, and install a variety of energy conservation devices and systems (Fig. 1.7). For residences, these devices include storm windows, weatherstripping, insulation jackets for water heaters, and low-flow showerheads. Contractors are available to perform more substantial housing energy conservation retrofits. In addition, it is now possible to buy energy efficient air conditioners, furnaces, and other appliances that consume significantly less energy than the typical models in the existing stock.

Consider, for example, the energy performance of top mount, automatic-defrost refrigerator-freezers (the most popular style). The typical model now sold consumes 40% less electricity, and the best new model consumes 55% less electricity than a typical 1972 refrigerator. Furthermore, consumption levels of highly efficient prototypes, custom-made models, and foreign models are down another 35 to 80% compared to today's average new model (Goldstein 1984; Geller 1985). Improvements in insulation, motor-compressors, and general product design are responsible for these impressive developments.

The story of refrigerators is not exceptional. Great progress has been made in designing and building homes that require very small "auxiliary" (i.e., fuel or electricity derived) space heating (so-called superinsulated housing). In the commercial sector, computerized control systems are being installed in large buildings to monitor and manage the building's energy-use systems to reduce energy costs without adversely affecting the comfort of building occupants. Also, numerous energy efficient lighting products have been introduced in recent years.

As we discuss in Chapters 9, 10, and 13, the great challenge lies in disseminating these new, cost-effective technologies on a massive scale. In this regard, some promising signs are appearing in the private sector. For example, energy service companies (ESCOs) are providing a boost for the implementation of conservation measures in larger commercial and apartment buildings. As of 1984, more than 100 firms were arranging financing and installing energy-efficient equipment for their clients; the number of energy service companies is expanding rapidly (Klepper 1984). ESCOs commonly provide an engineering audit; select appropriate equipment; and purchase, install, and possibly even operate the equipment. In some cases, the operating cost savings achieved because of the investment is shared between the building owner and the ESCO.

Fig. 1.7. Advertisements for energy-efficiency products and services.

2

WHY PURSUE FURTHER ENERGY
CONSERVATION IN BUILDINGS?

The significant gains in energy efficiency that occurred during the past decade, the host of new energy efficient products and design techniques, and the array of private and public sector programs to stimulate efficiency improvements might be interpreted to mean that nothing more needs to be done. That conclusion could not be further from the truth.

A. ENERGY COSTS

Fuel costs have become and continue to be a major concern. Consumers in the U.S. paid $420 billion for energy in 1984. Furthermore, energy costs rose from 8% of total Gross National Product in 1970 to 14% in 1981 and then declined to 11% in 1984 (Geller 1985). The "price shock" traumas of 1973 and 1979 have been transformed into the malaise of routinely high prices in the 1980s. This is particularly true for fuel oil and natural gas, the major energy forms used to heat buildings. Average residential gas and heating oil prices more than doubled in real terms from 1973 to 1983 while electricity prices increased 30% (Energy Information Administration 1985).

The trend towards increasing fuel prices is not expected to subside for either natural gas or electricity. A recent utility industry forecast shows residential electricity rates rising at an average rate of 2.3% per year above inflation from 1984 to 2000 (Electrical World 1984). With continuing deregulation, natural gas prices may increase as well.

Lower income families have been hit particularly hard by increasing energy prices. While the average household spent only 4% of its income on household energy in 1980, low-income households spent 24% of total income (Fig. 2.1). Furthermore, home energy expenditures as a fraction of income more than doubled in low-income households from 1972 to 1980 (Cooper et al. 1983).

The Federal government funds both fuel assistance and weatherization programs for low-income households. Unfortunately, almost 90% of this money is used for fuel assistance, leaving insufficient funds to address the roots of the problem (Williams et al. 1983). A large-scale, subsidized program to upgrade the building shells and heating equipment in low-income homes would return the energy bills of the poor to manageable levels and reduce government fuel bill assistance.

Routinely high energy prices have a host of other adverse impacts. One recent assessment showed that, as of 1980, energy accounted for an average of 31% of the operating cost in rental housing, up from 14% in 1973 (Cooper et al. 1983). This price rise can cut profits for landlords and purchasing power among tenants, yielding a general decline in the quality of rental housing (Cooper et al. 1983). Rental units are being abandoned, especially in large cities, because landlords find them unprofitable. Retrofitting these multifamily buildings would lower operating costs and make these buildings more attractive to owners and renters.

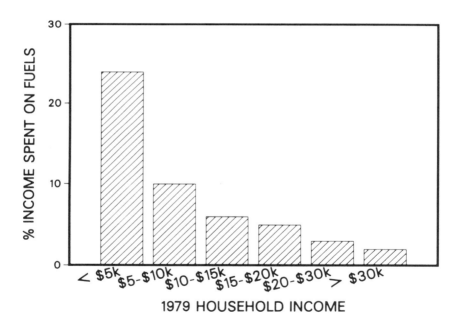

Fig. 2.1. Relationship between residential fuel expenditures and income (Energy Information Administration 1984). Low-income households pay a large fraction of their income for fuel bills.

High energy costs affect owners and managers of commercial buildings as well as households. The 1984 Building Owners Poll showed that 66% of the owners of commercial buildings rated energy costs as their top problem (National Real Estate Investor 1984). According to this survey, 93% of the respondents believe conservation is important and 41% believe that conservation has become increasingly important.

B. FURTHER SAVINGS POTENTIALS

The opportunities for large additional increases in energy efficiency and, more important, in dollar savings for consumers have been well documented. For example, the National Audubon Society (1984) published a proposed national energy plan that would cut total energy use in residential and commercial buildings by 30% and save consumers $93 billion (in 1983 dollars) in the year 2000. The plan calls for substantial capital investment in energy efficiency in buildings between 1985 and 2000. An annualized capital investment of $47 billion yields reductions in fuel bills double that ($93 billion), clearly reflecting enormous untapped conservation potential in our nation's building stock.

An earlier study by the Solar Energy Research Institute (SERI 1981) concluded that energy use in residential and commercial buildings could be cut almost in half in the year 2000 with no loss in amenities. This estimate was based on conservation measures that were technologically and economically feasible in 1980.

Another assessment showed that residential end-use energy demand in the U.S. could decline by one-half from 1980 to 2020 if efficiency improvements are implemented to the maximum extent (Williams 1985). Likewise, commercial sector end-use energy demand could drop by one-third in spite of commercial floor area increasing by one-half. This forecast is based on technologies that are cost-effective and that are either commercially available or under development.

With the opportunities for massive conservation in buildings identified at the national level, a few local and regional energy planning organizations have set out to "capture" some of this potential. The City of Austin, Texas, released a conservation plan in 1984: a five-year program to cut the electricity demand of the city by 550 MW. This decrease is a 20% reduction relative to forecast power demand without stepped-up conservation.

The Northwest Conservation and Electric Power Plan (Northwest Power Planning Council 1983) includes a set of vigorous conservation programs designed to reduce annual electricity use in residential and commercial buildings by 20% by the year 2002, relative to usage in which current efficiencies are maintained. The programs required to yield these savings cost, on average, less than 2 cents/kWh saved, far below the 4 cents/kWh break-even point (Fig. 2.2). According to the Northwest Power Planning Council, "In this twenty-year plan, conservation is the cheapest resource for the [Pacific Northwest] region and it will play a major role in meeting future electric energy needs."

These examples indicate that major institutional commitments to conservation are being made based on current savings opportunities. However, the savings potential will be even larger than indicated above because of the introduction of new technologies in the future. A group at ORNL (Spiewak et al. 1983) estimated that additional R&D could cut building energy use in the year 2020 by 50% relative to what it would be without that additional R&D, a net benefit (in 1980 dollars at a 3% real

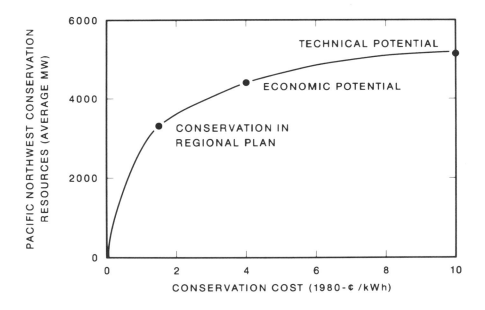

Fig. 2.2. Electricity conservation supply curve for residential and commercial buildings in the Pacific Northwest. The 3300 MW included in the Regional Council's plan is equivalent to the output of four large power plants (Northwest Power Planning Council 1983).

interest rate) of $900 billion. Nearly half the potential benefit of improved technologies would come from changes in building envelopes; about one-third from improvements in space heating systems; and the remainder from improvements in appliances, water heating equipment, and lighting systems.

C. OTHER BENEFITS

In addition to the obvious energy savings and economic advantages, a number of secondary benefits or "externalities" are associated with energy conservation in buildings.

Some forms of energy conservation in buildings are highly labor intensive and provide much greater employment opportunities than alternative investments in energy supply systems. For example, an evaluation of the federal weatherization program showed that 36 direct jobs and 15 supporting jobs are created for each $1 million invested annually in weatherization (Garey and Stevenson 1983). In contrast, each $1 million per year spent on operating a new coal-fired power plant is associated with only about five jobs (at the plant and in coal mines).

The economy as a whole stands to gain from improving energy efficiency. First, cost-effective conservation investments free up capital that would otherwise be spent on power plants, oil wells, etc. These energy supply facilities already account for 39% of total U.S. capital expenditures on new plant and equipment (Williams 1985). Second, the U.S. energy trade deficit (the value of energy imports minus exports) was almost $50 billion in 1984 (Energy Information Administration 1985), almost half the merchandise trade deficit that year.

In 1984, 30% of the petroleum used in the U.S. was imported. The U.S., Europe, and Japan are still heavily dependent on Middle Eastern oil supplies. By lessening our dependence on imported oil supplies, the aggressive pursuit of energy conservation will contribute to improved national security. Reducing oil use in residential and commercial buildings by approximately 30% in 1984 could have displaced nearly half the oil imported from Arab OPEC nations (Energy Information Administration 1984).

So far, the official U.S. response to potential attacks on oil tankers and facilities in the Persian Gulf has been to prepare for a military intervention to protect our national interest. However, a "marketplace intervention" to ensure that our demand for oil products is reduced would be a much safer and more secure response (Kelly 1983).

Although fuel supplies are abundant today, underlying physical resource limitations still exist. That is, supplies of petroleum, natural gas, and coal are all finite; improved energy efficiency postpones shortages of nonrenewable resources and provides more time during which alternative energy resources can be developed.

By decreasing the need for conventional energy supplies, vigorously pursuing energy conservation can reduce such environmental problems as air pollution and, in particular, acid rain. Also, without major reductions in world coal use through conservation and other means, atmospheric CO_2 levels will dramatically rise by the end of the next century (Goldenberg et al. 1985), which might cause significant climatic changes.

If all domestic refrigerators in the U.S. were as efficient as the top-rated models now being sold here, the need for at least ten large (1000-MW) electric power plants could be displaced. Such a reduction in power demand would ameliorate problems of nuclear waste disposal, reactor safety, air pollution, and fly ash disposal.

Only a few studies attempt to quantify some of the indirect benefits of energy conservation. Sebold, Thayer, and Hageman (1983) analyzed various conservation programs operated by San Diego Gas and Electric Company. Their analysis considered benefits related to environmental quality, dependence on foreign sources of petroleum, electric system reliability, comfort, and employment. These indirect benefits ranged from 25% to more than 40% of the fuel bill reduction (direct benefit) for programs that saved electricity. For programs that saved natural gas, the indirect benefit was smaller, ranging from 5% to 10% of direct benefits.

Finally, the U.S., with about 5% of the world's population, accounts for 25% of its commercial energy consumption. By pursuing conservation to the technical and economic limit, the U.S. would remedy some of this imbalance, help keep world fuel prices lower, and thereby increase worldwide economic growth. Moreover, aggressive action on the part of the U.S. could represent a big step towards long-term energy security and sustainability for the world (Goldenberg et al. 1985). Dealing with our energy problems in a farsighted, rational manner would also provide valuable lessons for other resource issues involving supply constraints (e.g., water, timber, mineral, and wildlife resources).

3

UNDERSTANDING THE BUILDING CONTEXT

A. INTRODUCTION

Assessment of energy efficiency of buildings must acknowledge the overall context of building design, financing, construction, occupancy, and operation. This involves a complex interweaving of people, professions, institutions, materials, money, decisions, regulations, and environmental factors. This chapter sketches some of these interactions and the context for energy use in buildings.

B. CLIMATE AND COMFORT

Buildings provide the temperature, humidity, and lighting necessary for people to live and work productively and in comfort: buildings create a functional environment. In recent times, this environment has been provided through sophisticated technological devices, such as electric lighting and HVAC equipment. The resulting building forms require energy to operate these comfort related technologies.

The particular aspects of climate that are of interest to the building designer are temperature, wind, humidity, insolation, and precipitation. Analysis of the local climate will suggest to an energy-conscious designer what the principal comfort and discomfort conditions are and what design strategies to use. Buildings in different regions will have different shapes, materials, orientations, envelope characteristics, and earth—ground relationships. In cold climates, the design might use the sun as much as possible for winter heating, while buildings in temperate zones may have much more open interior configurations. Architectural features desirable in homes in different regions are summarized in Fig. 3.1.

The American Society of Heating, Refrigerating and Air-Conditioning Engineers (ASHRAE) publishes a psychrometric chart, which defines a general human comfort zone (ASHRAE 1977). This comfort zone includes a range in dry bulb temperature of 70 to 80°F and a range in

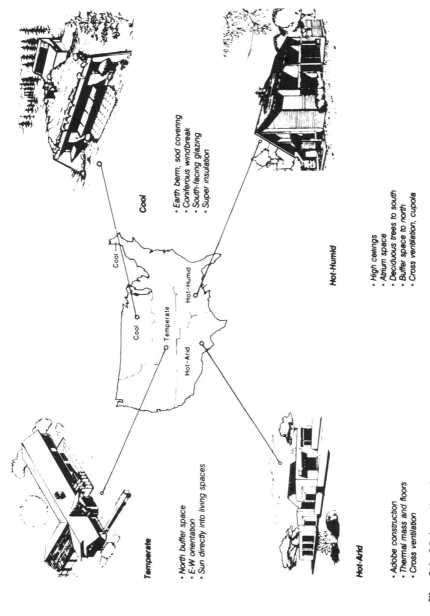

Cool

· *Earth berm; sod covering*
· *Coniferous windbreak*
· *South-facing glazing*
· *Super insulation*

Hot-Humid

· *High ceilings*
· *Atrium space*
· *Deciduous trees to south*
· *Buffer space to north*
· *Cross ventilation, cupola*

Temperate

· *North buffer space*
· *E-W orientation*
· *Sun directly into living spaces*

Hot-Arid

· *Adobe construction*
· *Thermal mass and floors*
· *Cross ventilation*

Fig. 3.1. Major climatic regions of the U.S. and housing designs appropriate to each region (American Institute of Architects 1982).

relative humidity from 20 to 80%. As humidity increases, the maximum comfortable dry bulb temperature decreases.

If the dry bulb temperature and the humidity fall within that zone, 70% of the population will feel comfortable. However, a particular person may not experience thermal comfort even within this zone. Age, sex, health, activity patterns, and other factors may affect comfort.

Examinations of outdoor conditions (i.e., plots of temperatures and humidity levels superimposed on the psychrometric chart) are used to determine appropriate design strategies (Watson 1979). For example, ambient conditions below the comfort zone suggest the need for humidification, while conditions to the right of the zone suggest the need for sensible cooling (Fig. 3.2). External temperatures just to the left of the zone suggest that passive solar is an appropriate strategy, while more extreme conditions further to the left suggest active solar and/or conventional heating systems (Givoni and Milne 1981).

Even under the best circumstances, a general HVAC system can only partially provide human comfort within a building. The physiological state of an individual may place that person in thermal stress even though measurements within a building indicate that the person should be comfortable. Peculiar cold air movements, uncontrolled cold or hot interior surfaces, equipment radiating heat towards a person, mean radiant temperature conditions created by large glass areas, clothing variations, and nearby human activities and behavior (smoking and odors) can stress building occupants. Consider the following:

- One person's need for fresh air, satisfied by opening a window, may be another person's discomfort because of drafts.

- A person who opens the blinds to sit in warm sunshine may create a glare problem for another.

- Space far from an exterior window may be thermally comfortable, but the region near the window may be uncomfortable because of radiative cooling.

- An adult standing in a room may be comfortable, while a child nearer a cold floor may be uncomfortable.

Providing comfort is more than installing heating, cooling, lighting, and ventilation systems. These systems need to be flexible and responsive to occupant needs and allow people to open and close windows, modify temperature and humidity conditions, adjust illumination levels and light quality, and control the flow of fresh air.

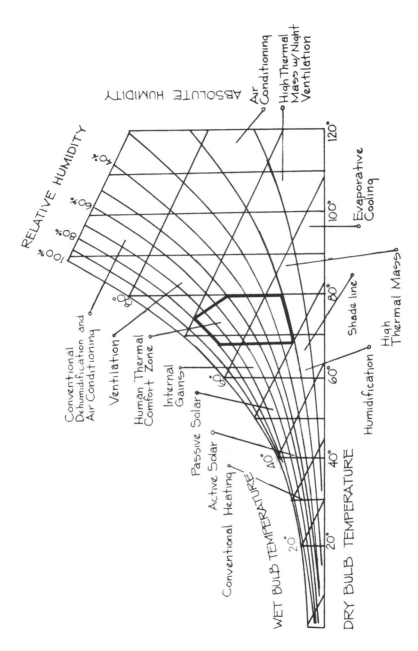

Fig. 3.2. The psychrometric chart, showing the relationships between wet and dry bulb temperatures and absolute and relative humidity (Watson 1979). The chart also shows design strategies suitable for various environmental conditions.

Finally, we have to live with the objects we place in the natural environment. The aesthetic (visual) qualities are important. "Tacked-on" technologies, such as solar collectors on the roof, may be energy efficient, but they can be an eyesore if designed without visual sensitivity. Reflective glass may limit thermal gains into a building, but its blinding effect can create both a safety problem and discomfort for the people across the street. Being able to see the view, hear the sounds of nature, and maintain some control over one's environment are important psychological attributes of a well-designed building.

C. THE URBAN CONTEXT

The comfort and energy aspects of a building are influenced by the urban context in which it is set: the building density, street layout, types of buildings, location, orientation to sun and wind, and layout of communities. Urban planning and design affect the pattern of sun and shade on buildings of various heights; the degree of protection from rain, dust, and wind; and ventilation conditions around and within buildings. Dimensional aspects, such as the height of structures, spacing between buildings, and the homogeneity or variety of building sizes and heights, also influence comfort and therefore building design.

Building dimensions are critical in terms of ventilation. A building standing in an open field creates a wind shadow on its leeward side. The wind velocity there is much less than in front of the building. Lining buildings up behind each other reduces the air velocity around the downwind buildings. As a result, wind velocity in a built-up area is generally much lower than in open country.

Streets and other open spaces enable passage of wind among the buildings and improve ventilation conditions. Single buildings projecting above neighboring ones may modify patterns and velocity of air flow near the ground.

Substantial microclimatic differences occur between a city and its surrounding suburbs and rural areas. For example, temperatures in the city may be several degrees higher than in the country because cities contain large thermal masses in their buildings and receive waste-heat discharges from those buildings. On the other hand, access to the sun during winter months is minimal in many urban areas. These differences between urban and rural areas can be assets or liabilities, depending on the comfort and energy strategies chosen for individual buildings.

D. BUILDING FORM

At each phase of the design and construction process, energy saving decisions are possible (Table 3.1). The greatest effect on energy savings can be realized during the schematic design phase when the basic decisions about building design are made.

The space-conditioning loads of a building result from a combination of internal and external heat gains and losses. Each gain or loss is produced by different sources and is affected by different parameters. External heat gains or losses are those caused by the external environment (temperature and solar radiation). They can be reduced through shading and increased thermal resistance. Internal heat gains (i.e., lights, people, and equipment) are more difficult to reduce because they are produced within the building's interior. Because residential buildings generally have large surface-area-to-volume ratios and small internal loads, their energy use is dominated by the external environment and characteristics of the building shell; such buildings are called skin-dominated. Large commercial buildings, on the other hand, have small surface areas relative to their internal volumes and have large internal loads. Hence, their energy use is dominated by activities inside the building and they are called internal load dominated.

The relative amount of internal heat gain versus external heat gain or loss is primarily a function of building form and envelope characteristics. Variations in building shape greatly affect the amount of exterior surface area for a given volume enclosed (Fig. 3.3). A tall narrow building has a high surface-area-to-volume ratio. It has a small roof and is affected less by solar gain on this surface during summer months. On the other hand, tall buildings are generally subjected to higher wind velocities and thus have greater infiltration rates and heat loss. Low buildings have a greater roof area relative to wall area, so special attention must be given to the roof's thermal characteristics.

A study of surface-area-to-volume ratios indicates that elongation along the east-west direction is optimal in all climates (Olgyay 1963). For commercial buildings, the principal penalty of a north-south axis is increased operating costs because of higher peak cooling loads. The amount of elongation needed depends on the climate. In cool and hot-dry climates, a compact building form exposing minimal surface area to a harsh environment is desirable. In temperate climates, more options exist that do not have severe drawbacks. In hot and humid climates, buildings should be elongated liberally in the east-west direction because of intense summer solar radiation on the east and west sides (Fig. 3.4).

Table 3.1. **Energy savings possible at various stages in the building design process**

Stage in design process	Potential energy savings possible
Program predesign	0–10%
Schematic design	40–50%
Design development	30–40%
Construction documents	0–10%
Construction management	0–10%
Post-construction	10–20%

Source: American Institute of Architects (1981).

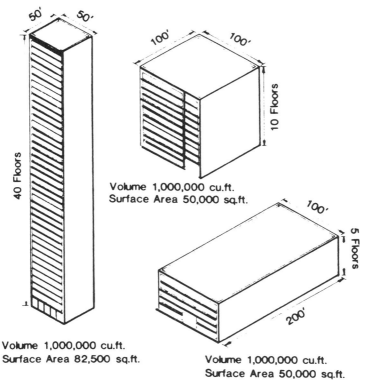

Fig. 3.3. **Variations in building surface area with different building forms (American Institute of Architects 1982).**

Buildings can be located on the ground, underground, partially in the ground, or above the ground plane. Below the frostline, underground temperatures remain stable at approximately 56°F. In climates with severe winter or summer temperatures, underground or in-ground construction provides temperate outside design temperature, as well as reduced infiltration.

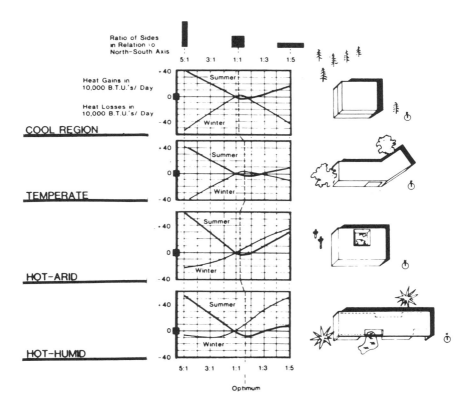

Fig. 3.4. The effects of building shape on heating and cooling loads in different regions (Olgyay 1963).

E. BUILDING ENVELOPE

The building envelope is the boundary between the building and outside conditions. It filters light, affects air and thermal flows into and out of the building, contributes structural strength, provides fire protection and safety, and gives access.

The envelope includes four subsystems: roof, wall, floor, and earth contact. These subsystems are made up of numerous components and materials and may incorporate other systems, such as HVAC, plumbing, and electrical services. The three major factors determining heat flow into and out of a building are external conditions (temperature, wind speed, and insolation), the area of the building exposed to the elements, and the heat transmission of the exposed areas. Building envelopes can store thermal energy (Fig. 3.5) or create a thermal lag regardless of whether that energy comes from the interior or exterior. A wall with high thermal capacitance can significantly delay the transfer of thermal energy from the exterior to the interior, or it can function as a cooling source for the interior.

Glazing plays a particularly important role in building envelopes, including:

- Providing direct sunlight when desired,

- Providing daylight,

- Providing view and visual contact with the outdoors,

- Providing natural ventilation for direct physiological comfort,

- Providing cooling of the structure or interior through natural ventilation,

- Providing intrinsic aesthetic values.

It is generally advisable to have small windows because they better prevent indoor temperature changes. However, with special design details, such as insulated shutters, large windows can provide thermal advantages (Givoni, 1984). Other options for windows and skylights (e.g., layers of glazing, heat-absorbing glass, and reflective glass) also affect overall energy performance.

Color is another characteristic of building envelopes that affects energy use. The colors of the roof and walls determine the solar absorptivity, the solar energy that is naturally absorbed by the building's envelope, which in turn affects heat flow into the interior of the building and indoor temperatures.

F. BUILDING INTERIORS

In large commercial buildings, energy consumption is primarily determined by the building's interior: electric lights, equipment, people, and other contents. Internal loads vary widely among various types of buildings (Table 3.2).

Partitions affect energy consumption by reducing daylighting (increasing lighting costs) and by blocking natural ventilation (requiring mechanical ventilation). Where natural ventilation for cooling can freely move from the building's windward side to the leeward side, cooling loads can be reduced. Partitioning is most often found in offices where the building perimeter is private office space separated from the core of the building.

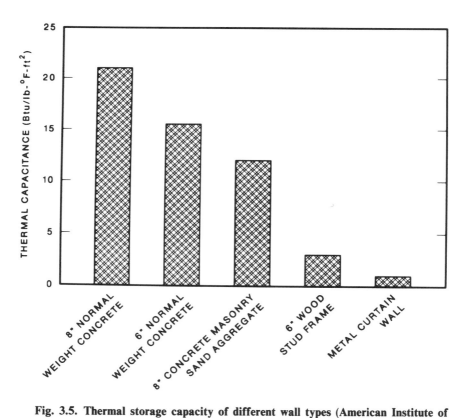

Fig. 3.5. Thermal storage capacity of different wall types (American Institute of Architects 1982).

Table 3.2. Typical energy budgets for different building types

Building type	Percentage of total energy use, by function		
	Envelope[a]	Contents[b]	Lighting
School	15	45	40
Housing	50	30	20
Office	10	40	50
Hospital	10	40	50
Commercial	15	45	40

[a]Envelope includes walls, windows, roof, and floors.
[b]Contents include ventilation, appliances, elevators, motors, fans, and other miscellaneous loads.
Source: American Institute of Architects (1981).

Typical floor plans assume that a building benefits from daylighting to a depth of 15 feet. The remainder of the interior must use electric lighting, which increases internal-heat gains. A building where most of the floor area is adjacent to the building envelope can have a lower lighting power requirement than one that does not. Rectangular configurations with a small central core are preferred (Fig. 3.6). Conditions at or near a window can improve daylight penetration. For example, light-shelves allow deeper penetration of daylight.

One of the internal load components is the heat generated by people. Human metabolic (heat generation) rates vary from about 400 Btu/hour for sedentary activities to up to 900 Btu/hour for level walking and 3000 Btu/hour for very heavy work (Givoni 1976). This energy is radiated from the body to heat sinks within the interior, and it raises air temperatures. In addition, moisture evaporates from the body, raising indoor humidity levels and placing a dehumidification load on the HVAC system.

G. PASSIVE ENERGY SYSTEMS

The microclimate presents the designer with assets and liabilities vis-a-vis environmental comfort and energy. Passive energy systems, which use these assets for heating, cooling, and/or daylighting by natural means, reduce the amount of fuel required by a building. Passive energy systems are intrinsic to the building (Givoni 1984), and they involve the selection of site, location, orientation, floor plan, circulation patterns, window placement, and building materials.

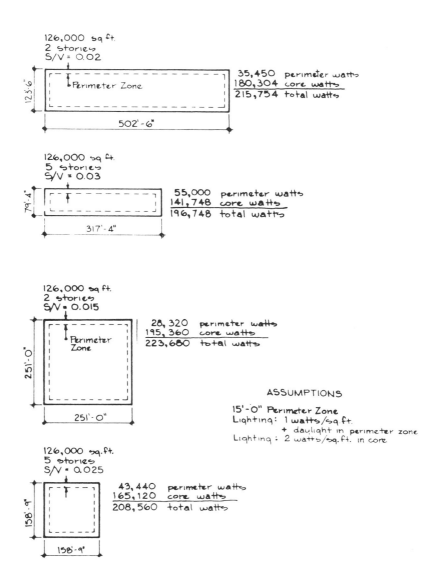

Fig. 3.6. The relationship between building shape and electric power requirements for lighting (American Institute of Architects 1982).

Passive solar heating systems typically consist of south-facing glass (the collector), thermal storage (absorber and/or mass), the occupied space itself, transport systems, and energy flow controls. The collector admits solar radiation onto the absorber; the absorber converts radiation to heat; the thermal mass holds heat not immediately used; various forms of distribution methods (convection, conduction, or radiation) transfer heat between the thermal mass and the space; and the energy flow controls reduce heat loss through the collector. Controls also provide shading, ventilation, and/or increase the amount of radiation transferred to the collector.

Passive cooling systems can include several different strategies (Givoni 1986). Ventilation can be used to cool people and/or structural components of the building. Radiant cooling involves heat transfer from building components to the sky through longwave radiation. Evaporative cooling (suitable in hot dry climates) is accomplished by evaporating water. Finally, heat can be transferred to the earth surrounding the building, for example by drawing ventilation air from the crawlspace of a house.

Daylighting is another passive option. Vegetation can be used to filter sunlight, reduce glare, or otherwise make sunlight usable for visual tasks in buildings. The surroundings (adjacent buildings, ponds, or other objects) increase or decrease daylight.

H. BUILDING EQUIPMENT

Building equipment includes HVAC equipment, lights, appliances, fans, pumps, elevators, controls and other components. There can be many equipment options to serve a particular function, and just as many factors (fuel costs, building use, load conditions, etc.) that influence equipment choice. Many innovations in building equipment and their significance for energy use are discussed in Chapters 6 and 7.

I. OCCUPANCY AND BUILDING USE

Building occupants obviously have large effects on energy use. For example, if occupants lower the indoor temperature setting by 2°F, they can cut annual space heating use in a single-family home by 10%. The number of people in a building and the amount of time spent in the building also affect energy use. Building occupancy, use, and duration are important design parameters that determine not only the comfort requirements but also the potential thermal energy available from the interior.

Office buildings, churches, schools, theaters, conference centers, lobbies, and malls are not used 24 hours a day. While empty, these buildings continue to consume energy to protect the building and its equipment and to reduce the time required to bring a space back to occupancy temperatures. Buildings whose occupancy can be extended through multiple uses are theoretically more efficient. Institutions that serve multiple functions may not find it necessary to construct yet more buildings. Rather, the space problem might be solved through rescheduling of activities and space management.

Another efficient energy-design potential is thermal-energy exchange. Buildings whose occupancy generates heat during the day can be integrated with environments primarily occupied during the night (such as hotels). The excess heat from one can then be used by the other on a diurnal cycle. Multiuse buildings, such as the 100-story John Hancock Center in Chicago, combine offices, housing, shopping, and cultural facilities. Properly designed HVAC systems within such a building can take advantage of thermal energy sharing between offices and housing.

J. EXISTING BUILDINGS VERSUS NEW BUILDINGS

Interest in improving the energy efficiency of existing buildings is high because in any given year the number of new buildings constructed is very small compared with the stock of existing buildings. A small increase in the efficiency of existing buildings can, therefore, have a greater impact than a big increase in the efficiency of new buildings.

The potentials of redesign and retrofit of existing buildings include the following strategies:

- Upgrade HVAC, lighting, appliances, and other building equipment.

- Reduce the amount of interior partitioning to allow deeper penetration of daylight, thermal balancing, and increased cooling through natural ventilation.

- Add insulation to the building envelope and increase thermal resistance of glazing.

- Insulate pipes and ducts.

- Limit smoking to designated areas to reduce ventilation requirements.

- Modify occupancy schedules according to seasonal patterns (e.g., by occupying a building later in the day in the winter when the sun has had a chance to heat the building and by occupying that building earlier in the day in the summer).

- Paint exteriors a color appropriate for heating or cooling priorities.

- Capture internal heat for storage or for transfer to heat-loss areas.

- Install devices on the exterior to shade the wall or roof where heat gain is a problem.

K. INSTITUTIONAL CONTEXT

The design and use of buildings are influenced by a complex set of institutional elements, discussed in Chapter 8. Decisions are often clouded by conflicting attitudes reflecting cultural, financial, political, psychological, and aesthetic biases. This confusion is caused by all those who influence the design process: individual (or corporate or government) clients; members of the design/engineering professions; financial institutions; producers of materials; contractors who employ craftspersons; trades, and union people who construct buildings; standard setting and regulatory agencies; trade or professional associations; professional journals; and public opinion.

Material usage, structural requirements, and safety requirements are prescribed by building codes. Codes differ from community to community and from state to state, and the modification of building codes is accomplished through a variety of methods and legislative processes. The primary purpose of building codes is protection of the general health, safety, and welfare. Most building codes are derived from one of four model codes developed by Building Officials and Code Administrators, International; International Conference of Building Officials; National Conference of States on Building Codes and Standards, Inc.; and Southern Building Code Congress, International.

Lending institutions have their own building standards. Some of these incorporate Federal Housing Administration Minimum Property Standards to take advantage of government sponsored insurance programs.

Construction and assembly of buildings is influenced by a fragmented building-trade structure. Union jurisdictions establish work procedures and the amount of prefabricated building components that can be used. Those procedures are set through a complex network of contracts and agreements.

Many industries establish performance standards for their own products or assemblies. Lighting level recommendations by the Illuminating Engineering Society, ventilation and heating standards by ASHRAE, and material standards by the American Society for Testing and Materials are perhaps the most familiar. In addition, the building industry has standards for the handicapped, environmental impact, noise control, and workmanship.

The basic purpose of local zoning is to establish and maintain high quality neighborhoods and communities. For example, in a residential district, zoning prohibits commercial uses, establishes lot size requirements, and sets densities. Patterns of building bulk, orientation, and configuration result from constraints imposed by zoning laws.

Conditions that affect building energy use can be shaped by utility companies through rates, requirements, or inducements, such as more favorable rates for well-insulated buildings, lower rates during off-peak hours, and a variety of energy conservation services; see Chapter 10.

State public service commissions influence the amount and kind of energy used in their jurisdictions. They can determine such factors as the fuels available, the rate structure, and contruction of power plants. In addition, state regulatory commissions can, and do, order their utilities to implement various energy conservation programs. Generally, the basis for these orders is evidence that conservation resources are less expensive than is construction of new energy generation facilities.

Legislation creating new energy related government agencies with new authorities was passed at both the state and federal levels following the 1973 oil embargo. The DOE affects energy use in buildings through the research it sponsors and the programs it manages. Most state governments, through their state energy offices, also operate energy conservation programs. Some of these programs are operated in response to federal legislation and funding; others are operated at the state's initiative. The programs conducted by these governmental agencies are described in Chapter 9.

L. BUILDING OWNERSHIP

Building owners influence design decisions in both new and retrofit construction. The owner's motivation to have a building that is energy efficient depends, in part, on whether or not that owner has to absorb the cost of energy or whether it can be passed on to the occupant. Another important element is whether the owner or occupant benefits directly from qualitative improvements in the building (such as comfort) and whether this qualitative improvement increases the productivity of the occupants.

Not surprisingly, an Office of Technology Assessment study (1982) showed that owner-occupants (corporations, individuals or small businesses, condominiums, and public institutions) are more likely to retrofit their buildings than are investor-owners (Chapters 5 and 7). Propensity to retrofit is also a function of building type. For example, master metered multifamily buildings are more likely to be retrofitted, regardless of ownership type, than individually metered apartment houses.

II

PROGRESS TO DATE:
WHAT WE NOW KNOW ABOUT BUILDING
ENERGY EFFICIENCY

As discussed earlier, we have two main objectives in writing this book. The first is to present a state-of-the-art review of current knowledge on energy efficiency in buildings. The second is to suggest an agenda for future research and programs to further improve energy efficiency. This part discusses what we know about energy efficiency in buildings and programs to stimulate efficiency improvements.

As we reviewed papers presented at the 1984 Santa Cruz conference; journal articles; government, national laboratory, and university reports; discussions with peers throughout the building energy efficiency professions; and our own experiences and knowledge, we were struck by the wealth of information produced during the past decade. Clearly, this book cannot provide comprehensive coverage of current knowledge on the subject. Instead, we present highlights of the major findings and offer many references for interested readers.

We consider this review important for two reasons. First, the explosion of information—technical, behavioral, programmatic—on energy consumption and efficiency in buildings cries out for organization and compilation. To some extent, the first section of the report from the 1982 Santa Cruz conference, *What Works* (Harris and Blumstein 1984), provides comparable coverage. This section updates and expands on the ideas and topics covered in *What Works*. Equally important, we believe, is the need to set the stage for our proposed agenda for future research and programs to further improve energy efficiency in buildings. Without a firm basis in current knowledge, it is difficult to formulate a credible plan for future activity.

This part of the book includes eight chapters. The first four are organized around traditional sector lines. Chapters 4 and 5 treat new and existing residential buildings, respectively. Chapter 6 discusses residential appliances and HVAC equipment. Chapter 7 deals with new and existing commercial buildings. The fact that we have three chapters on residential buildings and their equipment but only one on commercial buildings reflects the imbalance in attention historically paid to these two sectors.

Chapter 8 discusses human behavior: how people affect the design, construction, operation, and use of buildings. Chapters 9 and 10 describe the history and current status of government and utility programs intended to improve the energy efficiency of buildings in the U.S. The final chapter discusses indoor air quality.

4

NEW RESIDENTIAL BUILDINGS

A. INTRODUCTION

Approximately two million new residences are constructed each year in the U.S. Although this number represents only a small fraction of the roughly 80 million housing units in existence, new construction is important for two reasons. First, residential structures have very long lifetimes, on the order of a century. (Mobile homes, of course, have much shorter lifetimes of roughly 20 years.) Second, incorporating energy efficiency features into a new structure is much less expensive than retrofitting an existing structure. Also, fewer efficiency improvements are possible in existing homes.

The energy efficiency of new homes involves technical issues related to the design and construction of homes, the building shell, the heating and cooling equipment, the appliances installed within the building, and interactions among these energy systems. The behaviors of individuals and groups are as important as the technical aspects of energy efficiency. These behaviors include decisions made by architects, builders, financial institutions, realtors, and the ultimate occupants of new houses (see Chapter 8).

Government and utility programs help overcome barriers to increased energy efficiency. Such programs include the provision of information intended to influence the energy related decisions of various actors (e.g., rating systems that tell potential buyers what the energy efficiency of a new building is); standards that set minimum levels of energy efficiency for new construction; financial incentives (generally from utilities) that reflect the reduction of loads placed on the utility system by energy efficient buildings; and more-generous mortgages for energy efficient homes with larger allowed monthly payments because of lower energy operating costs. These programs are discussed in Chapters 9 and 10.

Analysis of the designs and construction practices used in 1975 and 1976 to build new houses showed substantial variation in energy efficiency (American Institute of Architects Research Corporation 1978). For example, the energy efficiency (in Btu/ft^2) of single-family homes varied by more than a factor of ten; the range among the middle 60% of these designs was more than a factor of two, from 34 to 85 kBtu/ft^2 (Table 4.1). This large spread suggests that substantial room exists for improvement in the energy efficiency of new residences.

Information on the actual aggregate energy performance of new homes (analogous, say, to the EPA fuel-economy figures for automobiles) is simply not available. Meyers and Schipper (1984) examined data collected by the National Association of Home Builders and show that the thermal performance of a new home shell increased substantially since 1973. The average R-value of attics in new homes increased from 14 in 1973 to 25 in 1981, and the percentage of windows with multiple glazings increased from 40% to 62%. Recent data from the NAHB show further improvements between 1981 and 1983.

Table 4.1. Design energy use for space heating and cooling for new homes constructed in 1975 and 1976

Building type	Number of buildings	Average size (ft^2)	Design energy performance (kBtu/ft^2–year)					
			Min.	20%	50%	Mean	80%	Max.
Multifamily low rise	307	849	10	22	31	43	57	123
Single Family attached	298	1368	13	26	34	47	63	128
Single Family detached	3886	1861	12	34	48	69	85	221
Mobile Home	102	1030	14	53	82	75	96	117

Source: American Institute of Architects Research Corporation (1978).

B. SINGLE-FAMILY HOMES

ACTUAL PERFORMANCE

Data collected and analyzed at Lawrence Berkeley Laboratory (Busch et al. 1984) show that it is feasible to construct "low-energy" homes that use considerably less energy for space heating than the average new construction. Williams et al. (1983) showed that the average thermal integrity factor (winter heat loss through the building shell per unit floor area and heating degree day, Btu/ft^2-HDD) for new low-energy homes constructed in Minnesota in the early 1980s was about one-third the value for all existing U.S. homes and about two-fifths the value for homes constructed in 1980 (Fig. 4.1). And finally, data on 208 low-energy single-family homes showed them to have an average thermal integrity of 2.5 Btu/ft^2-HDD, compared with 5.0 for 1979 new construction practices and 8.0 for the current U.S. residential building stock (Ribot et al. 1984). Clearly, the construction of homes that use very little fuel for space heating is feasible.

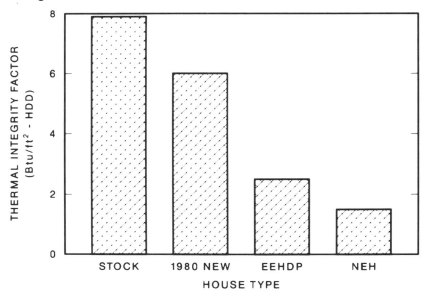

Fig. 4.1. Thermal integrity factor for alternative residential building envelopes (Williams, Dutt, and Geller 1983). Stock refers to existing (1979) single-family homes; 1980 New refers to homes built in 1980; EEHDP refers to home built as part of the Minnesota Energy Efficient Housing Demonstration Program; and NEH is the Northern Energy Home.

The data analyzed by Busch and et al. (1984) show that energy use for space heating can be reduced in several ways (Table 4.2). These alternatives (many of which are complementary rather than mutually exclusive) include superinsulation, passive solar, active solar, earth sheltering, and the use of high efficiency heating equipment. Superinsulated and passive solar homes are the dominant types of low-energy homes.

Superinsulated homes are those with very low values of thermal conductance for the building shell. This low conductance typically results from large amounts of insulation in ceilings, walls, and floors plus triple-pane windows. For example, the Northern Energy Home (a manufactured low-energy home from New England) includes 8 in. of polystyrene insulation in the walls and roof, yielding an insulation value of R-38. Compare this R value with those for the average of new homes constructed in 1983: R-28 in the roof and R-13 in the walls.

Passive solar heating designs involve the careful use of large south-facing windows, increased thermal mass, and control systems. The windows allow solar energy to penetrate the building shell during the day, and the thermal mass permits storage of this energy for use during cloudy days and at night. Such designs often include manually or automatically controlled window covers to reduce conduction and radiative heat losses from the house at night.

Low-energy designs usually involve extensive efforts to reduce infiltration of outside air. Typically, a continuous polyethylene air/vapor barrier is installed on the inside of all exterior walls. Lischkoff and Lstiburek (1984) discuss an alternative method involving conventional

Table 4.2. Summary of energy performance results for new low-energy homes

Category	Number in data base	k-value[a] (Btu/h–°F)	Balance temperature (°F)	Cost of conserved energy	
				elec. homes ($/MBtu)	gas homes ($/MBtu)
All homes	319	355	55	7.8	5.0
Superinsulated	196	276	59	5.8	3.9
Passive solar	197	249	50	5.6	4.2
Active solar	26	461	57	15.9	0.0
Earth sheltered	9	219	50	0.0	0.0

[a]The k-value, a measure of the rate of heat loss from a building, includes infiltration.
Source: Busch et al. (1984).

drywall that they suggest is simpler and less expensive to install. Many designs with low infiltration levels (e.g., less than 0.25 air change per hour) include an air-to-air heat exchanger to provide sufficient ventilation for the occupants. Indoor air quality issues associated with these "tight" houses are discussed in Chapter 11.

Low-energy designs can differ markedly in their sophistication. At one extreme, are "brute force" designs that simply apply more and more insulation to a conventional house. Others attempt to carefully integrate the various components of the house with each other and with the external environment. In Sweden, for example, factory methods that achieve very close tolerances between physical elements are used to produce efficient, high quality, affordable homes (Schipper, Meyers, and Kelly, 1985). Swedish home building techniques are starting to be used in the U.S.

Most of the low-energy homes constructed to date are in the northern parts of the U.S. and Canada. Williams et al. (1983) computed the likely space heating thermal performance for three types of homes in six cities throughout the U.S. (Fig. 4.2). They considered the typical 1980 new house, houses from the Minnesota Energy Efficient Housing Demonstration Project (EEHDP), and the Northern Energy Home (NEH). Their calculations suggest that the Minnesota and NEH designs would yield substantial energy savings in every region of the U.S. In mild climates (e.g., Fresno, California) these low-energy homes are essentially zero-energy homes. Thus, internal loads (the energy deposited within the conditioned space by people, appliances, and solar gains) are especially important in low-energy homes. Corum (1984) discusses the tradeoff between improved appliance efficiency and space heating and air conditioning loads. In the Pacific Northwest, with substantial heating loads and almost no cooling loads, the energy savings from high efficiency appliances is partially offset by greater energy use for space heating. In hot climates (e.g., Florida and Arizona), the net effect of high efficiency appliances is likely to be an even greater energy saving because of the reduced cooling loads.

A second conclusion one can draw from Fig. 4.2 is that energy use for space heating could become a minor energy use in homes. Prior to the early 1970s, space heating accounted for more than half of the total energy use in a typical home. Ten years later, space heating accounted for only 40 to 45% of the typical home's energy use. Design, construction, and use of low-energy homes may further increase the relative importance of water heating, refrigerators, and other energy uses currently considered "minor."

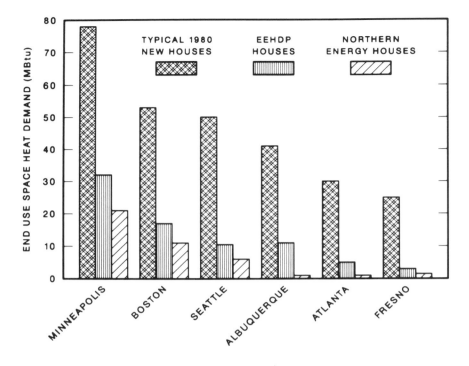

Fig. 4.2. Space heating energy use for a 1500 ft² single-family house with alternative envelopes in various climates (Williams, Dutt, and Geller 1983).

DESIGN OF HOMES IN HOT CLIMATES

Considerable research, development, and dissemination of techniques have been carried out for greatly reducing residential energy consumption in colder parts of the U.S. (e.g., superinsulated housing). However, little attention has been devoted to the problems of constructing and operating energy efficient housing in hotter areas. This situation is unfortunate for a number of reasons. First, in spite of the milder climate, energy end use per household in the Southern part of the U.S. is only slightly below the national average (Energy Information Administration 1984). Furthermore, household consumption per heating degree day is 50% greater in the South than it is in the nation as a whole. Finally, new construction is now centered in the "Sun Belt"; in 1983, the Southern region accounted for 860,000 housing starts out of a national total of 1.7 million.

The need to provide substantial amounts of both space heating and space cooling throughout the year greatly complicates the design task in hot regions. For example, well-insulated buildings that rely heavily on internal loads for space heat can present severe cooling problems in the summer (Fairey 1983). Also, many hotter areas are also humid, with large latent cooling loads.

Techniques for avoiding heat gain in housing in hotter climates are available (Fairey 1983). Solar gains can be limited through building orientation, shading windows, protecting the roof and east- and west-facing walls, and installing radiant barriers (aluminum foil) in attics. Radiant barriers can reduce heat transfer from the attic into the living space by up to 44% (Fairey 1982); additional work is underway to evaluate their use in walls.

Techniques for naturally removing heat from buildings have also been developed (Givoni 1981). Where the air is dry, evaporative cooling can be employed. If the air is cool during the night (under 68°F), buildings with high thermal capacitance can save energy if they are "flushed" with outside air. This strategy works best if the building is shaded from heat gain during the day. Other passive-cooling techniques include use of the thermal capacitance of the ground and radiation to the night sky (Clark and Burdahl 1980).

The use of high thermal mass in warm, humid climates is controversial (Fairey 1983). If thermal mass is used, protection of the mass from heat sources (during the summer) and exposure to heat sinks (during the winter) are recognized design principles, and techniques for achieving these objectives are under development.

Moisture is of concern in both the winter and summer in hot, humid climates. Care must be taken in the use of vapor barriers as well as in the selection of air conditioners with adequate latent cooling capablity. Good building ventilation is also desirable. Both natural and forced ventilation have been studied, and guidelines for home designers and builders are available (Chandra, Fairey, and Houston 1983). Although several studies of moisture problems in buildings are underway, moisture transport processes are not yet well understood (Cleary 1984).

ECONOMICS OF IMPROVED ENERGY EFFICIENCY

Granted that low-energy homes offer substantial energy savings relative to conventional construction, one must then ask whether the increased costs of the new designs are justified by reductions in fuel bills. Busch et al. (1984) examined the economic performance of the low-energy homes in their data base and concluded that almost all such homes are cost-effective

(Fig. 4.3 and Table 4.2). In part, low-energy homes are economically attractive because the higher cost of the building shell is partially offset by the reduced cost of the heating equipment. In some cases, it may be possible to completely eliminate space heating systems and rely solely on solar and internal heat gains.

It is important to emphasize that not all low-energy home designs are cost-effective. A case in point is earth-sheltered housing, which was reviewed by Wendt (1982). He noted that although space heating energy use for a well-designed earth-sheltered home might be only one-fourth that used for a conventional new home, the substantially higher cost of earth-sheltered homes cannot be justified everywhere. Ideal locations for earth-sheltered housing include those with substantial temperature extremes, low

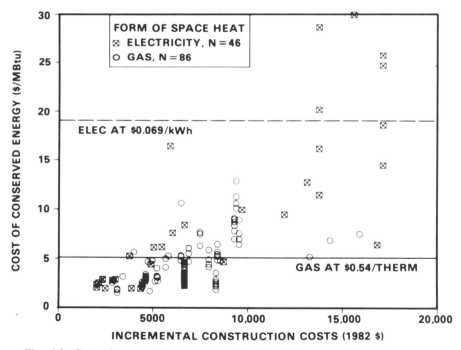

Fig. 4.3. Cost of conserved energy as a function of incremental contractor cost for 132 low-energy homes (Busch et al. 1984). Points that fall below the appropriate fuel price line are cost-effective; i.e., the higher cost of efficient construction is more than offset by lower energy operating costs.

humidity, appropriate subsurface soil conditions (little moisture, good drainage), and the presence of steeply sloped ridges running in a northeast/southwest direction.

Active solar homes are often not cost-effective at today's fuel prices. Also, low-energy designs intended to reduce space heating energy use are probably not economical in southern parts of the U.S., where air conditioning is a more important energy use than space heating is.

Ribot et al. (1982) concluded that superinsulated homes are almost always cost-effective, that passive solar homes are marginally cost-effective, and that active solar houses are not. LBL's 1984 update (Busch et al. 1984), however, shows that passive solar homes are cost-effective. Because the price of electricity is so much higher than the price of natural gas, most low-energy designs are more cost-effective for electrically heated homes than for gas heated homes.

BARRIERS TO CONSTRUCTION OF LOW-ENERGY HOMES

Although the design and construction of very energy efficient homes are both feasible and cost-effective from the viewpoint of the "economically rational" person, conventional construction practices lag far behind for many reasons. The obstacles to more efficient homes include imperfect information (neither builders nor homebuyers have the information needed to make cost-minimizing decisions), constrained capital availability (homebuyers seek to minimize their down payments with little regard for monthly costs for fuels), and differences in perspective between the supplier and buyer of the house. In the last situation, the builder seeks to minimize initial cost to increase the likelihood of a quick sale, while the intelligent homeowner would like to minimize the lifecycle costs of owning and operating the home (Blumstein et al. 1980).

Some buyers of new homes, however, might reasonably avoid high efficiency homes and purchase conventional homes if they are unsure about how long they will stay in their new home and uncertain about the extent to which the higher initial costs will be reflected in the resale value of the house (Holt 1984).

Analysis of single-family construction practices by O'Neal et al. (1981) supports the preceding concerns about underinvestment in thermal efficiency of new homes. They computed the discount rates implied by purchases of energy conservation options in new homes. Although the discount rates vary with heating fuel choice, location, and assumptions about future fuel prices and mortgage financing, the implicit discount rates were generally much higher than market interest rates. For example, assuming that the house is purchased with a 25-year mortgage with an interest rate 3% higher than inflation and that future fuel prices increase at the same rate as inflation (i.e., constant "real" fuel prices), the average (across the ten cities considered) discount rates are 30% for gas heated homes, 51% for oil heated homes, and 74% for electrically heated homes. If the market for energy efficiency in new homes worked well, implicit discount rates would be close to actual interest rates (5 to 10%).

Laquatra (1984) conducted an economic analysis of the sale price of 81 new homes constructed in Minnesota as part of the Energy Efficient Housing Demonstration Project (see Hutchinson et al. 1984). His results show that sale price increased by $2500 per unit decrease in the thermal integrity factor (measured in Btu/ft^2-HDD). This increase in sale price was economically attractive to the homeowner over a 30-year lifetime but not over a 6-year lifetime. In other words, if the household does not move for 30 years or is able to sell the house at a price that reflects the thermal-performance improvement, then the increased initial investment is justified. However, if the household moves after only a few years and is unable to increase the subsequent sale price to recoup the initial investment, then the thermal efficiency improvements are not economically attractive to the original homeowner.

C. MULTIFAMILY HOMES

PREDICTED PERFORMANCE

About half the multifamily units are in high-rise apartment buildings with five or more units. High-rise buildings, which can be as large as 50,000 ft^2 and more, share some characteristics with larger commercial buildings. Small multifamily units (low-rise) share many characteristics with single-family homes.

Detailed analysis of energy efficient designs for new multifamily buildings, sponsored by DOE in the mid-1970s, yielded a wealth of data on current design practices, their relationship to existing standards, and the potential for improvement; see Table 4.3. Relative to typical mid-1970s design practice, energy efficient designs for both high- and low-rise buildings were expected to cut energy use by almost one-third.

The calculated energy consumption of multifamily buildings ranged widely across climate regions in the United States. For example, the highest energy budget for a multifamily high-rise is 58 $kBtu/ft^2$ for Minneapolis, Minnesota; the lowest is 30 $kBtu/ft^2$ for Los Angeles and San Diego, California.

Much less information is available on the performance and economics of new multfamily buildings than on single-family homes. For example, not one of the dozen papers on multifamily buildings presented at the 1984 Santa Cruz Summer Study dealt with new buildings.

NEW LOW-RISE MULTIFAMILY HOUSING

Low-rise housing units use less energy for heating per unit floor area than their single-family counterparts use. Small passive solar systems can make a large contribution when combined with heat from people, lights, and equipment. Shared walls cut heating energy consumption, but they also can create poor natural light, and any increase in internal load is more difficult to reduce in summer because of limitations with natural ventilation. Finally, uncertainty about tenants' commitment to energy conservation makes it riskier to invest in certain components, such as moveable insulation, that require user participation.

The Massachusetts MultiFamily Passive Solar Program (MFPS) is frequently cited as a successful low-rise housing project (Rouse 1983). Under this program, the state housing authority provided $2.2 million in construction awards for 30 housing projects (750 apartments). The building types included townhouses, rowhouses, and mid-rise buildings with the primary focus on housing for the elderly. The projects included many energy conservation features and a wide range of passive solar systems: direct-gain windows, trombe walls, and sunspaces. On the average, the MFPS sponsored apartments are expected to use only 40% as much energy for space heating as similar projects that did not receive assistance and that used conventional design strategies.

Table 4.3. Comparisons of predicted site energy use by multifamily buildings of various design

Building type	Predicted energy use (kBtu/ft^2)		
	Mid-1970s design practice	Design meeting HUD standards	Energy-efficient design
High-rise	63	56	43
Low-rise	52	48	36

Source: American Institute of Architects (1978).

The project experience indicates that:

- Passive solar principles are simple and do not place major restrictions on the design process.

- Passive solar design is an effective energy-saving strategy for multifamily buildings. Added conservation and solar features cut energy use for space heating substantially below that of apartments conforming to the Massachusetts energy code.

- Passive solar buildings are affordable; added costs for conservation and passive solar features are often less than added costs for other special design elements.

- HVAC systems were initially overdesigned by as much as 200%. Mechanical ventilation systems that ensure good air quality should be integrated with ducted air-to-air heat exchangers to reclaim waste heat from the ventilation exhaust.

NEW HIGH-RISE MULTIFAMILY HOUSING

Multifamily high-rise buildings have been out of the mainstream of energy research. Although such buildings are important, they are hybrid buildings. As income properties they tend to be managed like commercial buildings, yet they share many characteristics with single-family dwellings, such as their occupancy schedule and nighttime use intensity.

In comparison with low-rise buildings, high-rise units have higher energy consumption for heating and lighting, but about the same energy consumption for other purposes. High-rise residences have less opportunity to take advantage of passive solar gains because they have less envelope exposure per unit of floorspace.

One comprehensive analysis of the effects of design on the energy use of multifamily buildings was the BEPS program. Twenty-five existing multifamily buildings ranging from 17,000 to 282,000 ft^2 and located throughout the U.S. were analyzed and redesigned to determine how their energy use patterns could be reduced. Redesign of the multifamily buildings typically produced an overall 30% energy savings (Table 4.3). Following are some of the conclusions about design effects.

Site and Orientation

Fifty-five percent of all multifamily units are located in central cities, where siting and access to sun and wind are critical concerns. Although project economics dictate the number and type of units required for successful marketing and zoning codes often determine density, buildings can frequently be reoriented to take better advantage of solar access and to gain better control over external heat gains and losses.

Solar Access

Passive solar heating is possible if the building faces south. Internal and external shading can be used to control insolation. The fact that multifamily buildings are skin-dominated can be overcome by the use of an atrium to form a thermal buffer space. Even with increased glazing to the south, total glass area can be reduced, and double or triple glazing can be employed, depending on the climate.

Conduction

Form compaction can minimize conductive surface area in northern climates. Conduction is critical not only because of the skin-dominated building type but also because of the high level of occupancy at night, when outside temperatures are lowest. As a result, multifamily buildings experience a higher temperature differential (inside to outside) for longer periods than do other buildings. For energy savings, night setback would minimize this differential. Whether or not occupants use such night setbacks is, of course, a problem unless automatic controls are employed.

High-rise buildings tend to have less wall insulation than low-rise multifamily buildings. One reason is that low-rise buildings generally use wall framing construction, which produces a larger cavity (for insulation) between the inner and outer walls than that between masonry walls.

High-rise buildings can use common spaces (corridors, lobbies, etc.) and other unconditioned areas (stairwells, elevator shafts, storage rooms, etc.) as buffers to reduce the load on spaces where the temperature is allowed to float. This strategy decreases the amount of conditioned floor space.

Lighting

In high-rise buildings, compact configurations limit use of daylighting unless specific efforts are made to design the envelope for maximum daylighting. Light-shelves are seldom found in such buildings; their use needs further investigation. Switching to fluorescent fixtures in public spaces as well as in bathrooms and kitchens further improves energy efficiency.

Natural Ventilation

Virtually no data are available on the use of natural ventilation for cooling high-rise multifamily buildings. Natural ventilation is often difficult, given the double-loaded-corridor configuration. Designs that allow movement of air from the windward to the leeward side can take advantage of two types of cooling: convective cooling of the human body and radiative cooling (provided the building mass is at a lower temperature than the indoor air). High-rise buildings tend to have considerable thermal mass with concrete floors, masonry walls, and ceilings that can be effectively cooled with outdoor air.

As the foregoing indicates, there are many strategies for reducing energy use in high-rise buildings. But they are infrequently used because of institutional barriers, lack of clear and convincing demonstrations, and difficulties in obtaining money to pay for these improvements.

D. MOBILE HOMES

Mobile homes account for 10 to 15% of new housing construction but only 6% of existing housing (Bureau of the Census 1983). Thus, ways to improve energy efficiency in mobile homes, especially in new units, should be considered.

Analysis of alternative mobile-home designs by Hutchins and Hirst (1978) showed considerable potential to improve thermal performance (Fig. 4.4). They found that improving thermal performance beyond the levels required by the 1976 mobile home standards promulgated by the U.S. Department of Housing and Urban Development was cost-effective, even at the fuel prices prevailing in 1976.

Perhaps because the HUD standards are lax, because mobile homes are inexpensive, and because mobile homes are found primarily in regions with mild winters (the Sun Belt and the Pacific Northwest), the average thermal performance of new mobile homes is much worse than that of new single-family homes. The average thermal performance of the existing stock of mobile homes (not just new units) is almost 25% worse than that of single-family homes (Mills 1984). In addition, mobile home thermal performance is poor because of geometry, limitations that are imposed by the need to keep costs low, small wall cavities that prevent installation of sufficient insulation, and transportability requirements that prevent use of overhangs and large thermal mass.

Mills (1984) identified 39 high efficiency mobile homes in the U.S. and Scotland that had been monitored (Fig. 4.5). These homes use only about 20% of the space-heating energy required by mobile homes meeting the HUD standards. Although Mills does not discuss the economics of improving mobile home thermal performance, one side-by-side comparison of HUD-standard and identical homes with energy-saving features suggests that improved thermal performance is economically attractive. Improved mobile homes in Florida used 25% less energy than those with the HUD design and cost only $750 more (Mills 1984).

Improving thermal performance of new manufactured housing is very cost effective for consumers. Upgrading new manufactured homes from the current HUD mobile home standards (R-7 in walls and R-7 or R-14 in the ceiling) to the Farmers Home Administration standard for site built homes (R-19 in walls, R-30 in the ceiling in areas with 2500 to 6000 heating degree-days) would reduce space-heating bills about 50% in most areas of the country (Mineral Insulation Manufacturers Association 1984).

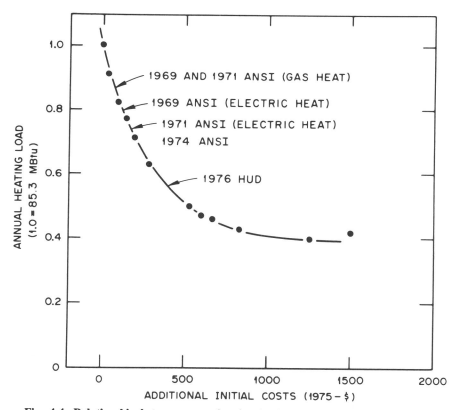

Fig. 4.4. **Relationship between space heating load and additional initial cost for a new mobile home in Kansas City (Hutchins and Hirst 1978). Each point on this conservation supply curve represents one or more efficiency improvements.**

The increased first cost associated with these improvements would be about \$650 and \$900 for single-wide and double-wide mobile homes, respectively. The payback period on the extra first cost would be less than three years in virtually all climate zones (Duke Power 1983; Housing Assistance Council 1984).

E. UNRESOLVED ISSUES

As the preceding discussions indicate, considerable knowledge about the design, construction, operation, and performance of space heating in residential structures exists. Clearly, we have come a long way during the past several years.

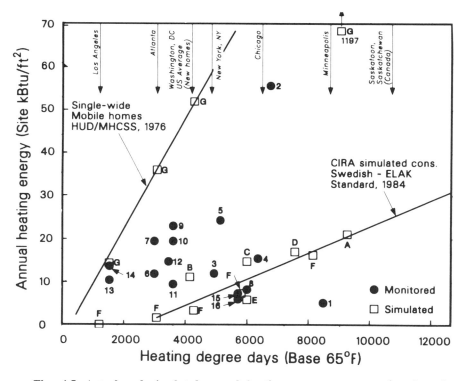

Fig. 4.5. Actual and simulated annual heating energy use as a function of heating degree-days in various types of manufactured homes (Mills 1984).

Not surprisingly, our understanding of these processess is far from perfect. Most of the activity, and therefore most of the available data, on low-energy homes comes from northern climates. Our understanding of the benefits and costs of low-energy designs in southern climates is meager. More information is needed on design options that reduce air-conditioning energy requirements in both dry and humid climates. Because most southern electric utilities experience their peak power demands in summer, both the energy savings and the peak-load reductions (kWh and kW) from improved building design must be measured. It may turn out that the major economic benefit of low-energy homes in the south would be their reduced peak-load requirements rather than their lower total energy requirements.

Information on multifamily and mobile homes is also scarce. Additional analysis of energy efficiency improvements and their costs is needed. But the greatest need is for empirical evidence concerning the performance and economics of various low-energy designs for these two dwelling types.

Current design tools are inadequate and too complicated for most builders to use, especially with respect to passive solar designs. Better and easier-to-use analytical tools are needed to determine the optimal type, location, and size of south-facing windows; the optimal relationship between glazing and shading (to reduce insolation during the summer); and the optimal relationship between glazing and internal thermal mass. Improved design tools are also needed to analyze different methods of cutting air-conditioning energy use. Those design tools need to consider both sensible and latent loads and both energy and peak-power reductions.

The effects of improved thermal performance on other attributes of a structure need further exploration. These needed investigations include consideration of moisture in the attic and other locations, indoor air quality, comfort, and interior noise levels.

Similarly, little is known about the durability of energy efficiency improvements. Because these improvements are expected to last for 20 to 30 years, their performance (and possible degradation) needs to be monitored for several years after construction.

More information is also needed on the behaviors of homeowners and organizations involved with the creation of new homes (architects, engineers, manufacturers, builders, realtors, and banks); see Chapter 8. In particular, little effort has been devoted to understanding the decisionmaking processes involved in the selection of a new home or the role of improved energy efficiency in those decisions.

Finally, given that most states now have some form of thermal performance standards for new building construction, it is shocking how little we know about the actual energy saving effects of these standards. Similarly, we know little about the effectiveness of rating systems, utility incentives to builders, and other methods to encourage construction of efficient homes; these topics are investigated more fully in Chapters 9 and 10. All of these topics are fertile ground for research.

5

EXISTING RESIDENTIAL BUILDINGS

A. INTRODUCTION AND REVIEW OF RECENT TRENDS

According to the 1980 Census, roughly two-thirds of the 80 million households in the U.S. live in single-family units, slightly more than one-fourth live in multifamily units, and the remainder live in mobile homes (Bureau of the Census 1982). About 80% of these housing units will still be in use at the turn of the century and at that time will account for more than half the total housing stock (Solar Energy Research Institute 1981). Thus, saving energy in existing homes is very important.

Various studies (e.g., Meier, Wright, and Rosenfeld 1983; Solar Energy Research Institute 1981) show the enormous *potential* for reducing energy use in existing homes. One technique used to illustrate this large potential is conservation supply curves. These curves show the amount of energy that can be "produced" through retrofits of existing homes. As Figs. 5.1 and 5.2 show, substantial energy conservation resources are waiting to be mined.

This chapter discusses how the building shell, heating and cooling equipment, appliances, and behavior of residents can be modified to achieve those energy savings. Again, the focus is on cutting space-heating energy use because little is known about the performance and economics of cutting air-conditioning energy use.

Information on residential retrofit actions during the past decade is scanty, largely because collecting reliable and relevant data (e.g., how many inches of what type of insulation were installed in how many attics) is extremely difficult.

An analysis of trends during the past decade (Crane 1984) suggests that much of the retrofit potential has not yet been realized: "The residential energy reductions that have been achieved to date are only a fraction of what can be accomplished ... households do not seem to be saving any more energy than would have occurred if each household had set the heating thermostat down by about 5° F."

Annual surveys have been conducted by the Energy Information Administration (1984) and the Bureau of the Census (Annual Housing Surveys). Of these, the EIA data are more useful because they deal specifically with energy conservation. The Census data are important because they are derived from a much larger sample.

The EIA data show some improvements in existing homes during the past several years. The percentage of single-family homes with attic insulation increased from 71% in 1974 to 77% in 1980, and the percentage with storm windows on all windows increased from 45 to 52% (Meyers and Schipper 1984). From 1978 to 1980, 12 million households added ceiling insulation and nine million added wall insulation.

Analysis of trends in residential energy use from 1972 to 1982 by Adams et al. (1984) suggests that improvements to the shells of both

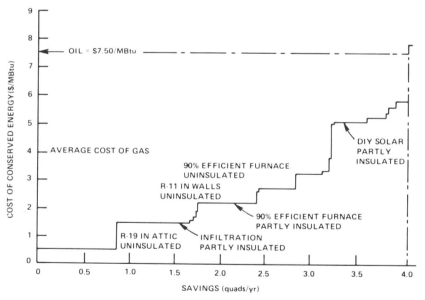

Fig. 5.1. Conservation supply curve for existing homes heated with gas or oil (Solar Energy Research Institute 1981). **The curve shows the energy saving available at different costs, where costs are amortized over the lifetime of the retrofit measures. The curve shows the conservation available in the year 2000; 1980 fossil-fuel heating use totaled 5.5 Quads.**

existing and new homes saved 0.7 QBtu in 1982, worth more than $4 billion in lower fuel bills. This is equivalent to 5% of residential energy use that year.

The most important factor in reduced household energy use since 1973 has been lower indoor temperatures (Meyers and Schipper 1984); surveys show that the percentage of households keeping their daytime temperatures at 70°F or higher declined from 85% in 1973 to less than 45% in 1981. Fels and Goldberg (1984) reached a similar conclusion. They analyzed natural gas sales data for New Jersey and found a 26% decrease in per household use of gas between the early 1970s and early 1980s. Almost half of this decline was attributable to changes in indoor temperatures, with the remainder caused by efficiency improvements. Additional information on household energy behavior appears in Chapter 8.

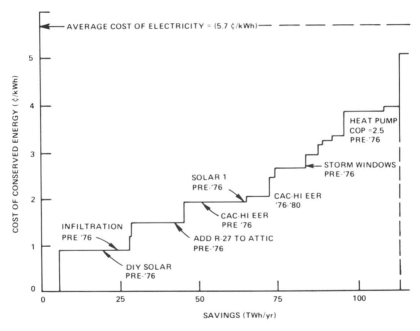

Fig. 5.2. Conservation supply curve for existing homes heated wtih electricity (Solar Energy Research Institute 1981). The curve shows the conservation available in the year 2000; 1980 electric heating use totaled 170 Twh.

B. SINGLE-FAMILY RESIDENCES

POTENTIAL AND ACTUAL SAVINGS

Goldman (1984) presents results from approximately 115 data sources on actual energy savings achieved in thousands of existing homes that received a variety of retrofits (Table 5.1). Most of the data are for single-family homes in the northern portions of the U.S.; little information was found on energy use for air conditioning. This national data base shows typical energy savings from retrofitting of 30 to 35 MBtu/yr, about 20 to 30% of the energy used for space heating in these buildings before retrofitting.

Traditionally, retrofits have focused on the building shell (especially attic insulation, with some attention to floor and wall insulation, storm windows and doors, and caulking and weatherstripping). Recent results (Crenshaw and Clark 1982; Kensill 1984; Gathers 1984; Proctor 1984) suggest that improvements to fossil fuel space heating systems may be more cost-effective than shell retrofits. For example, the National Bureau of Standards project on weatherization of low-income homes found that the payback period for houses that received both shell and equipment retrofits was less than one-half the payback period for houses that received only shell retrofits. Unfortunately, the two groups of homes were not comparable before retrofit. Moreover, neither the degree and quality of retrofit installation nor furnace efficiency were matched for the two groups.

Kensill (1984) analyzed oil-heated homes in Minnesota that were retrofit with a flame-retention burner, with typical building shell retrofit measures, or with both. Results showed that the burner retrofit yielded an average fuel-oil saving of 22%, the shell improvement 12%, and the combination of measures 29%.

Although results averaged over many houses demonstrate the energy saving effects of residential retrofits, the tremendous scatter across individual homes is disturbing. For example, comparison of actual natural gas savings with predicted savings in almost 350 retrofit homes in Minnesota shows only a weak relationship [$r = 0.33$; see Fig. 5.3; Hirst and Goeltz (1984)]. The reasons for this poor correlation are not fully known but probably relate to appropriateness of the particular retrofit measures installed, quality of retrofit materials and installation, and subsequent occupant behavior changes (e.g., changes in indoor-temperature settings).

Table 5.1. Key findings from current retrofit experience in residential buildings

	Utility programs	Low-income programs	Research studies	Multifamily buildings
Sample size	N = 19, 43730 homes	N = 30, 938 homes	N = 38, 352 homes	N = 28 bldgs.
Cost of retrofit (1983 $)				
median	705	1370	824	533
average[a]	1044 ± 702	1578 ± 863	1685 ± 2747	695 ± 551
Space heat savings (MBtu/yr)[b]				
median	38.4	30.5	27.8	15.1
average	40.3 ± 21.0	37.8 ± 26.2	34.3 ± 24.4	27.0 ± 27
Space heat savings (%)				
median	24	22	22	22
average	26 ± 11	24 ± 12	25 ± 14	26 ± 14
Simple payback time (yrs)				
median	5.7	9.2	6.4	4.7
average	10.3	11.4	9.5	7.9
Cost of cons. energy ($/MBtu)				
median	2.71	4.33	3.62	5.03
average	2.56 ± 1.29	6.33 ± 4.63	4.34 ± 4.05	5.26 ± 3.30
Real rate of return (%)				
median	25	6	17	11
average	23 ± 15	13 ± 14	31 ± 35	27 ± 31

[a]Mean ± standard deviation.
[b]Electric space heat savings are measured in primary energy units.
Source: Goldman (1984).

Goldman (1984) also notes a large variation in energy savings. In 50% of the retrofitted homes studied (those not in the top or bottom 25% in energy usage), energy savings varied approximately 70% from the median value. Dutt et al. (1982 and 1983) found similar variations in gas heated homes in New Jersey.

These large variations across individual homes suggest the need for better analytical methods to predict energy savings from retrofitting and for improved diagnostic tools to prescribe appropriate retrofits. Wagner (1984) reviewed the accuracy of 18 computer models of energy use with about 100 simulations. She concludes that the agreement between actual and predicted energy consumption depends on the quality of data available on building characteristics, occupant characteristics and behavior, and weather; on the user's skill in providing inputs to the model; and on the algorithm itself. Agreement between predicted and actual annual energy use was usually within ±20%. Agreement between predicted and actual energy *savings* is surely worse.

Recent improvements in diagnostic tools and energy auditor training should improve the accuracy of audit predictions. The House Doctor approach developed at Princeton University (Dutt et al. 1982) uses highly trained technicians, a blower door to identify air leakage sites, and portable infrared scanners to identify thermal anomalies (e.g., areas with insufficient insulation). In a controlled experiment, one set of homes was

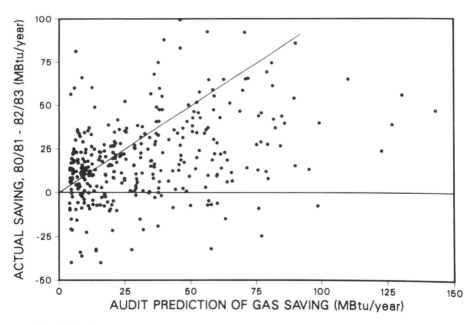

Fig. 5.3. Actual natural gas saving as a function of predicted saving for audited homes in Minnesota (Hirst and Goeltz 1984). The horizontal line separates homes in which energy use decreased after audits from those in which energy use increased. Actual and predicted savings are equal along the diagonal line.

visited by house doctors and another set of homes received both house doctor treatment and traditional major retrofits from contractors. The average natural gas saving for the 58 house doctor only houses averaged 15%; the 40 house doctor plus retrofit homes averaged 21%. The house doctor treatments were much more cost-effective than the major retrofits; the average payback period was seven years for house doctor treatments and 14 years for the major retrofits. These results suggest that well-trained technicians using sophisticated instruments can identify and implement inexpensive and very cost-effective retrofits.

The frequent disagreement between predicted and actual energy savings has substantial implications for the home energy audit programs operated by electric and gas utilities. These programs, frequently operated in response to the requirements of the federal Residential Conservation Service, include an onsite home energy inspection conducted by a trained energy auditor. Measurements made by the auditor are typically input to a computer program that simulates energy use (especially space heating) in the home. The computer program then estimates the energy savings that should be achieved with the installation of various retrofit measures.

Because of the high cost of conducting these audits and performing the calculations ($100 to $150 per home), because of poor predictive accuracy, and because fine-grained analysis is not always needed to prescribe appropriate actions or to persuade the household, some utilities have begun using simplified energy audits. In these audits, the absence of certain measures is sufficient evidence to recommend the installation of retrofits. In other words, recommendations for retrofits in individual homes are based largely on their effectiveness and cost-effectivness in typical (generic) situations.

On the other hand, the frequently poor agreement between actual and predicted energy savings after retrofit has led to the call for more research to identify, understand, and quantify the factors that cause these differences. For example, many homes could be retrofitted under closely monitored conditions. Detailed data could be collected on pre- and post-program energy consumption by end use, microclimatic conditions, the characteristics of the house, the measures installed, the quality of installation, and occupant behavior (especially selection of indoor temperatures). Careful analysis of these data might show the effects of these factors on actual energy savings and might indicate whether and how engineering heat-loss simulation models can be improved. Results might also suggest ways to better target homes for retrofit treatment and to select actions to be taken.

In addition to reducing household energy consumption, residential retrofit measures may yield additional benefits, such as improved comfort and increased insulation from exterior noise. For example, the addition of storm windows and insulation to the exterior of a house would increase the inside wall temperatures. Increased wall temperatures would, in turn, reduce radiative heat transfer from people to the walls, which would permit the occupants to be comfortable at lower indoor temperatures.

Alternatively, households may respond to the reduced fuel bills resulting from retrofitting by increasing their comfort and convenience (e.g., by heating previously closed rooms and/or raising indoor temperatures). Indeed, in one sample of 376 homes analyzed by Goldman (1984), energy consumption actually increased in 5% of the houses after retrofitting. And in Minnesota, natural gas consumption increased in 20% of retrofitted homes (Hirst and Goeltz 1984).

Residential retrofits might also have negative side effects. For example, adding insulation to an attic without ensuring that proper ventilation exists can lead to excess moisture in the attic and wood rot (Cleary 1984). Improper installation of insulation (e.g., over recessed lighting fixtures) can lead to fire hazards. Finally, house-tightening measures might reduce the rate of outside air exchange sufficiently to increase levels of indoor air pollutants (e.g., combustion products, radon, or chemicals emitted by furniture and carpets); see Chapter 11.

ECONOMICS OF RETROFIT

Goldman's (1984) review of national retrofit data shows that, on the average, most retrofits are cost-effective (Table 5.1). The contractor cost averaged $1400, and the energy saving averaged 36 MBtu/year across the 45,000 homes examined. The median payback period was six years for 19 utility-sponsored retrofit programs and nine years for 27 low-income weatherization programs.

Although residential retrofits are generally cost-effective, the large variation across households makes such investments risky for the *individual* homeowner. For example, Fig. 5.4 shows the electricity savings of homes retrofitted under the Bonneville Power Administration Residential Weatherization Pilot Program (Hirst, White, and Goeltz 1983). Those data display substantial scatter and indicate that these retrofits were not economically attractive in a substantial minority of the homes. Indeed, electricity usage increased after retrofitting in 12% of these homes.

This scatter in virtually all retrofit savings data is produced by a variety of influences. For example, changes in occupant behavior after retrofit complicate assessment of the economic attractiveness of residential retrofit. Careful examination of the BPA data (Fig. 5.4) suggests that some of the homes that showed very large changes in electricity usage after retrofit also substantially changed their use of wood for space heating.

Also, the lifetimes of the measures installed affect the economics of retrofits. Although most analyses assume that insulation measures last 20 to 30 years, no evidence exists on the actual long-term performance of these measures.

The economics of retrofitting depend strongly on whether measures are installed by a contractor or by the household itself. Certain products, such as caulk and weatherstripping, are very inexpensive but are labor-intensive to install. Some utilities seek to reduce the labor cost, both to the contractor and to the household, by subsidizing the cost. For example,

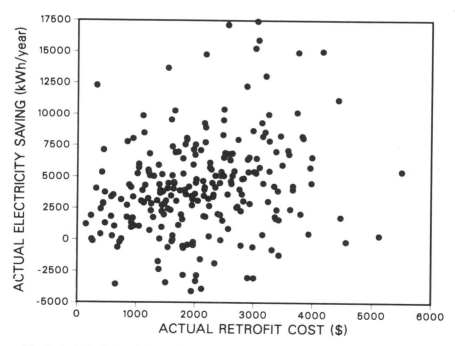

Fig. 5.4. Actual electricity saving as a function of actual retrofit costs for homes weatherized under the BPA pilot program in the Pacific Northwest (Hirst, White, and Goeltz 1983).

Florida Power and Light Company's Home Energy Loss Prevention (HELP) program pays 50% of the contractor cost for the installation of low-cost retrofit measures (e.g., outlet gaskets, weatherstripping, caulking, and water-heater wrap). Bundling the measures together reduces the contractor's labor costs because the measures are installed at one time, and the FP&L subsidy obviously cuts the household's cost in half.

Finally, the economics of retrofit depend on the perspective considered: that of the households receiving the retrofit measures, that of other households who might pay for these retrofits through higher utility bills or taxes, that of electric and gas utilities, or that of society as a whole. These perspectives differ because of differences in the value of the energy saving. Retrofit households gain benefits based on the average price of the particular fuel, society on the basis of the marginal cost of the fuel, and utilities (ratepayers in general) on the basis of the difference between marginal cost and average price. In addition, the perspectives differ depending on how the retrofit measures are paid (by a utility financial incentive, by a government tax credit, or by a household).

BARRIERS TO IMPROVED ENERGY EFFICIENCY

The factors that prevent implementation of known cost-effective retrofit measures and conservation practices are similar to those that inhibit construction of new low-energy homes. One obstacle is difficulty in obtaining site-specific, relevant, credible information on the likely benefits and costs of different actions for a particular home. This uncertainty is not surprising, given the large variation in actual effectiveness shown earlier. Typical homeowners may take little comfort from the fact that certain retrofit measures are cost-effective on the average. What they want to know is whether they will save energy and money in their particular home.

Kempton et al. (1984) investigated how households perceive and measure their residential energy use. They found that households seriously underestimate energy consumption for space and water heating and overestimate energy use for lights. This finding has significant implications for improving residential energy efficiency. If households are largely unaware of the major energy end uses in their house, they are unlikely to think of conservation measures associated with those end uses. Rather, they will focus on visible measures, such as turning off lights, that save much less energy than other measures.

Kempton and Montgomery (1982) also showed that households pay much more attention to their fuel bills (in dollars) than to energy consumption. Thus, the perception of cost-effectiveness for retrofit measures is seriously distorted during periods of rising fuel prices. That is, households who install retrofit measures and still see their fuel bills rise may view this as an indication that the measures were not effective.

One obstacle to retrofitting is uncertainty over its effect on the resale value of the house (Holt 1984). If households are unsure how long they will live in their present home and unsure about recovering their retrofit investment through a higher sale price, they are likely to avoid such investments. However, Johnson's (1981) study of the Knoxville, Tennessee, housing market showed that, holding other factors constant, a $1 saving in the annual fuel bill increases the sale price of the house by about $21. The implicit discount rates indicated by this relationship are quite low (on the order of a few percent, real), which suggests that the housing market operated "efficiently in capitalizing energy conservation investments." If these results can be generalized to other places and times, they suggest that housing markets adequately capture the value of retrofit investments. The apparent discrepancy between Johnson's finding concerning existing homes and those of O'Neal et al. (1981) suggests the need for more work to understand how households make energy-related decisions in their purchases of new and existing homes.

Thus, several factors hinder the homeowner in making a decision about upgrading the energy efficiency of a residence: uncertainty about what measures to take, uncertainty about the effectiveness of any measures taken, and uncertainty in locating a suitable contractor. This quandary is made particularly acute if, as some psychologists suggest (Stern and Aronson 1984), homeowners behave as problem avoiders, deferring action until absolutely necessary. The availability of "one stop shopping" services may overcome these barriers; see Chapters 9 and 10.

C. MULTIFAMILY RESIDENCES

POTENTIAL AND ACTUAL SAVINGS

Even less is known about energy use and conservation options for existing multifamily housing than for existing single-family homes. However, multifamily buildings deserve more attention than they have received because one-quarter of the nation's households live in these buildings. About 85% of the multifamily units in the U.S. are occupied by renters. In addition, these structures are disproportionately occupied by the poor and elderly (Bleviss and Gravitz 1984).

Most of the papers on multifamily homes presented at the 1984 Santa Cruz Summer Study discussed a variety of public and private mechanisms to encourage greater retrofit of these units. Only 3 of the 12 papers dealt explicitly with retrofit measures and practices. Interestingly, all 12 papers dealt with programs in the northern part of the U.S.; 6 of the 12 were from Minnesota.

The potential for energy efficiency improvements in existing multifamily buildings depends on many factors:

- Age of the building, its physical characteristics, and the type of HVAC equipment

- Building type (low-rise or high-rise and if low-rise whether the building is a townhouse, garden apartment, or other type)

- Form of ownership and occupancy

- Siting and orientation of the building in relation to solar access as well as degree of exposure to wind

- Quality of construction in terms of insulation, weatherstripping, and glazing

- Amount of public and semipublic spaces (lobbies, recreation areas, etc.)

- Floor-plan arrangements in relation to natural ventilation and daylighting

ECONOMICS OF RETROFIT

A review of the literature reveals a long list of retrofit options (Table 5.2). These retrofit options differ substantially in terms of their estimated energy saving, likely capital cost, and resulting cost-effectiveness (Table 5.3). For example, envelope-insulation measures tend to be expensive and less cost-effective than other measures, especially in large buildings. Heating and water-heating equipment controls, on the other hand, are very cost-effective. These observations suggest that the measures appropriate to multifamily buildings are often different from the measures best-suited for single-family homes.

LBL's BECA data base (Goldman 1984) includes information on energy savings and retrofit costs for 26 multifamily buildings. The average energy savings produced by retrofitting these 26 buildings, primarily resulting from reduced energy use in space heating, was 25 MBtu per apartment unit (Fig. 5.5). Most of the retrofits were cost-effective, especially those in gas heated buildings.

Some case studies highlight the energy efficiency potential for multifamily buildings. A pilot program was started in Massachusetts in 1982 (DaSilva and Waintroob 1984), the results of which were to be used

Table 5.2. Retrofit strategies for multifamily buildings

Retrofit strategy

ENVELOPE

Insulate exterior walls
Insulate roof or attic
Insulate floor or crawlspaces
Install storm windows and/or doors
Replace windows and doors
Weatherstripping and caulking
Convert south-facing windows to direct gain systems
Add thermal mass to direct gain systems
Install moveable insulation at windows
Convert windows to enhance natural ventilation
Paint exterior according to climate
Add exterior shading for cooling
Earth-berming where appropriate
Install entry locks
Add sunspace at south-facade

EQUIPMENT

Replace/modify hot water systems
Use low flow shower heads
New efficient heating system
New efficient cooling system
New heating/cooling controls
Install tank insulation
Replace incandescent light with fluorescent lights
Ventilate attic areas
Improve boiler efficiency
Insulate pipes and ducts
Install A.C. cover in winter
Positive shut-off for fireplaces
Install Casablanca fans

OPERATIONS

Reduce water temperatures
Lower daytime or nighttime temperature
Clean combustion air intakes
Recalibrate equipment and controls
Maintain pipes (leaks, etc.)

Table 5.3. **Sample list of retrofit options for multifamily buildings**

Retrofit	Category	Energy saving (kBtu/ft²)	Capital cost ($/MBtu-saved)
Low capital cost			
Roof spray	Envelope	15	Low (3)
Setback thermostats	Mechanical	7	Low (6)
Flow controls	Hot water	31	Low (0.5)
Insulate hot water storage	Hot water	34	Low (1)
Hot water vent damper	Hot water	8	Low (0.5)
Hot water heat pump	Hot water	40	Low (3)
Hybrid lamps	Lighting	15	Low (6)
Moderate capital cost			
Roof insulation	Envelope	7	Moderate (41)
Weatherstripping	Envelope	1	Moderate (39)
Window insulation	Envelope	8	Moderate (31)
Install heat pumps	Mechanical	22	Moderate (50)
Replace room air-conditioners	Mechanical	15	Moderate (26)
High capital cost			
Wall insulation	Envelope	27	High (81)

NOTE: Savings should not be added.
Source: Office of Technology Assessment (1982).

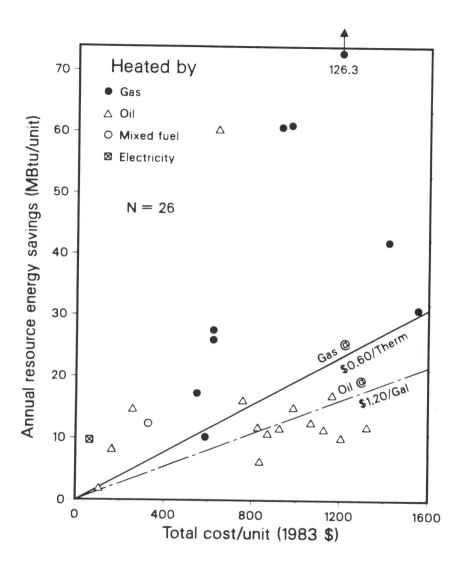

Fig. 5.5. **Actual energy savings as a function of retrofit cost for 26 multifamily buildings (Goldman 1984). Savings and cost are shown on a per apartment basis.**

to define technical, marketing, and cost parameters for a full-scale retrofit program. The program encompassed approximately 30,000 apartments (7.5% of the state's multifamily apartment population), 2700 buildings, and 24 million ft^2. The sample included townhouses, garden apartments, low-rise and high-rise multifamily structures. The most frequent recommendations from the energy audit included the following:

	Percentage of buildings in which measure recommended
Install low-flow showerheads and aerators	78
Reduce hot water temperatures	75
Lower daytime temperature	73
Install hot water tank insulation	71
Lower nighttime temperature	69
Replace incandescent lighting with fluorescent	47
Install ceiling insulation	42

If all these measures were implemented, it would cost $13 million and save an estimated $6 million in the first year; total gas usage might decrease by 40%, electricity usage by 40%, and oil usage by 35%.

A nonprofit student housing cooperative in Berkeley, California, undertook an innovative conservation program. Technical retrofits were combined with an energy education program for tenants and a change in energy billing procedures (Egel 1984). The results of the program included a 24% reduction in energy use (compared with no change in the control group), projected annual savings of more than $36,000, and a simple payback period of only 1.3 years (excluding the cost of labor, which was provided by co-op members).

The Berkeley project included training of student representatives from several buildings, followed by installation of retrofit measures by co-op members. Workshops on energy management were also provided for all residents. The retrofit strategies included low-cost measures as well as insulation and heating system modifications. The energy billing procedure was changed so that each building in the cooperative now receives and pays its own energy bill rather than having the bills paid by a central office. The central office assigns each building within the co-op an energy allotment. If a building saves energy, it "banks" part of that allotment. If the building wastes energy and exceeds its allotment, its residents pay the difference.

In a 159-unit public housing complex located in Trenton, New Jersey, a computer-based energy-management system was installed. The system uses remote temperature sensors located in selected apartments on each floor of each building and at one outdoor location. Readings taken on a 15-min cycle are fed into the computer. The computer controls heating system pumps and boilers to maintain comfortable temperatures in each apartment and to spot problems (Gold 1984). In the first year of the system's operation, weather-adjusted energy consumption was reduced by half, providing an annual savings of $55,000. Such a system, if commercialized, would cost about $65,000, which translates into a simple payback of slightly more than one year.

In the New Jersey project, some important lessons were learned about retrofits. First, it is unrealistic to reduce apartment temperatures below 75°F. Below this temperature, tenants complain, are uncomfortable, and use gas ovens for heating, a practice that is neither efficient nor safe. Some occupants judge comfort levels by whether the radiator feels warm, regardless of the actual temperature in the apartment. Because of this behavioral characteristic, the computer was programmed to keep radiators warm during the hours occupants were home and awake; temperatures were reduced during daytime working hours and at night. This particular retrofit project also points out that tenant and maintenance-staff education is critical.

BARRIERS TO IMPROVED ENERGY EFFICIENCY

Several barriers inhibit efforts to improve energy efficiency in existing multifamily structures. Perhaps the greatest obstacle is the "split incentive" between the tenant and the landlord. Because building ownership, payment of fuel bills, and building operation are split between the two parties, the benefits of efficiency improvements are also split. This situation is complicated by differences among buildings in who pays for fuels. In some buildings, especially all-electric apartment houses, tenants may pay the energy bills for their own apartments. In other cases, the building owner pays for the fuel that provides central space and water heating services, and the tenants pay their own electricity bills. Finally, in some cases the building owner pays for all energy costs.

The incentive to install energy conservation measures depends on more than who pays the fuel bill(s). Differences in owner type and building type also affect propensity to conserve, as shown in Tables 5.4 and 5.5 (Office of Technology Assessment 1982).

Frequent changes in both landlords and tenants further reduce the incentive for either party to retrofit. According to the 1980 Census, for example, almost half of the renters had lived in their present homes for only one year or less. Almost half of the homeowners, on the other hand, had lived in their present homes for ten years or longer.

High interest rates and the short term of typical loans further discourage building owners from investing in energy efficiency improvements (Bleviss and Gravitz 1984). In addition to these financial obstacles, the lack of reliable site specific information on the performance of retrofit measures makes it even more difficult for building owners to decide to install conservation measures.

D. MOBILE HOMES

According to Mills (1984), "Data on the technical and economic performance of manufactured home retrofits are rare." Perhaps this rarity is a consequence of the small percentage of homes in the U.S. that are manufactured. It may also be a consequence of the much greater difficulty in retrofitting mobile homes compared with single-family homes.

One effort to retrofit mobile homes in Minnesota showed smaller energy savings and longer payback periods than for retrofits of single-

Table 5.4. Typology of small multifamily buildings according to the likelihood of major improvement in energy efficiency

Owner type/ meter type	Building type	Likelihood of major improvement in energy efficiency	Owner's willingness to invest in retrofit
Owner-occupant	Frame	Moderate	Willing— low capital cost only
Owner-occupant	Masonry	Unlikely	Willing— low capital cost only
Absentee owner master-metered	Frame	Unlikely	Unwilling
Absentee owner tenant-metered	Frame	Unlikely	Very unwilling
Absentee owner tenant-metered	Masonry	Very unlikely	Very unwilling

Source: Office of Technology Assessment (1982).

family homes (Table 5.6). The poor performance of these mobile home retrofits may be partly caused by lack of knowledge concerning their effectiveness. The metal construction, thinner walls, unprotected floor, and other unique factors of mobile homes indicate that different retrofits are required for mobile homes than for single-family homes.

E. UNRESOLVED ISSUES

Many of the gaps in our knowledge of new homes apply also to our understanding of energy use in existing homes. In particular, the paucity of data concerning performance and costs of retrofits for multifamily and mobile homes is a major problem. Although our understanding of energy

Table 5.5. Typology of large multifamily buildings according to the likelihood of major improvement in energy efficiency

Owner type/ meter type	Building type	Likelihood of major improvement in energy efficiency	Owner's willingness to invest in retrofit
Institution master-metered	Central air or water system	Very likely	Very willing
Institution tenant-metered	Decentralized system	Likely	Willing
Condominium master-metered	Central air or water system	Likely	Willing— low capital cost only
Condominium tenant-metered	Decentralized system	Unlikely	Willing— low capital cost only
Individual or small partner- ship master-metered	Central air or water system	Moderate	Willing— low capital cost only
Individual or small partner- ship master-metered	Decentralized system	Very unlikely	Unwilling

Source: Office Technology Assessment (1982).

use in existing single-family homes is far from perfect, the attention devoted to this housing type is greater than its share of the total U.S. housing stock.

In addition, we lack sufficient data concerning appropriate retrofits for homes in the South, especially where both heating and air conditioning are major energy users. An important, related issue concerns the effects of retrofit measures on summer electric peaks. The load-shape impacts of retrofit measures are also important in areas with substantial electric space heating demands and winter-peaking utilities.

The substantial discrepancy between audit predictions and actual energy savings in individual homes suggests the need for careful monitoring to determine the factors that account for these differences. Results of such tests may indicate ways to improve energy audit predictions, retrofit techniques, and the training of auditors. The results may also show the importance of attention to detail in installation of measures.

The optimal combination of retrofits in individual houses is a subject of considerable discussion. "Optimal" refers to the most cost-effective combination of building shell and mechanical equipment improvements. Additional experiments are needed with different housing types, heating fuels, and climates to determine the most appropriate retrofit packages for given conditions.

Assumptions concerning the lifetime and long-term performance of retrofit measures (e.g., does attic insulation R value degrade over 30 years?) need to be tested in both laboratory and field settings. Economic calculations concerning the benefits and costs of retrofit measures depend strongly on these assumed lifetimes.

Finally, we lack adequate information on past and current trends in overall residential energy use. Although EIA's Residential Energy

Table 5.6. Comparison of retrofit performance in Minnesota:
mobile homes vs single-family homes

	Mobile homes (n = 35)	Single-family homes (n = 239)
Annual energy savings (MBtu/yr)	13	24
Energy savings (%)	10	14
Number of measures installed	3.2	4.7
Retrofit cost ($)	860	1040
Simple payback period (years)	16	11

Source: Mills (1984).

Consumption Survey provides a wealth of data, it is hard to interpret in terms of improvements in the technical efficiency of existing homes. Thus, we do not have an accurate picture of the extent to which behavioral changes vs efficiency improvements reduced residential energy use during the past decade. Nor is it clear how much of the potential for efficiency improvements (Fig. 5.1 and 5.2) has been achieved.

6

RESIDENTIAL APPLIANCES AND HVAC EQUIPMENT

A. TRENDS IN EFFICIENCY OF NEW PRODUCTS

Laboratory test procedures have been established by the National Bureau of Standards and DOE for measuring residential appliance efficiency (Government Printing Office 1981). The test procedures were developed for making the ratings that appear on appliance energy labels (see Chapter 9) and for tracking the average efficiency of new products over time. The ratings are also useful for other purposes, such as consumer education and determination of whether products qualify for rebates under utility incentive programs (see Chapters 9 and 10).

The available data on the average efficiency of new residential products since 1972 is shown in Table 6.1. The record for improving the efficiency of new products is mixed. For example, substantial gains were made in the average efficiency of new refrigerators and freezers during the 1970s, but only minor improvements have occurred since 1981. On the other hand, the average efficiency of new gas furnaces did not change during the 1970s but has risen since 1980. These results are a consequence of many factors including the availability, promotion, and acceptance of more-efficient models; the nature of purchasers and purchase decisions; and the regulatory and incentive programs used to stimulate adoption of efficient models. In general, products that show the largest efficiency gains are those for which the user most often makes the purchase decision and on which state standards have had the greatest impact (Geller 1984).

B. TECHNICAL DEVELOPMENTS

Despite the mixed record regarding improvements in the average efficiency of new products, tremendous progress has been made in developing and introducing highly efficient models. Table 6.2 compares the efficiencies of the top-rated models mass-produced as of 1984-85 and the efficiency of typical models sold in 1978 and 1983. Models substantially better than the average are available in all product categories.

Table 6.1. Trends in the efficiency of new products

Product	Efficiency parameter	Efficiency[a]					
		1972	1978	1980	1981	1982	1983
Gas furnace	Percentage seasonal efficiency[b]	63.2[c]	63.6	63.3[d]			69.6
Gas water heater	Percentage overall efficiencyb	47.4	48.2	47.9[d]			
Electric water heater	Percentage overall efficiency[b]	79.8	80.7	78.3[d]			
Central air conditioner	SEER[e]	6.66	6.99	7.60	7.83	8.31	8.43
Room air conditioner	EER[e]	6.22	6.75	7.02	7.06	7.14	7.29
Refrigerator/ freezer	Energy factor[f]	3.84	4.96	5.59	6.09	6.12	6.39
Freezer	Energy factor[f]	7.29	9.92	10.85	11.27	11.28	11.36

[a]Average efficiencies are weighted by manufacturers' shipments. Data provided by the industry associations AHAM, GAMA, and ARI. Also see "Consumer Products Efficiency Standards Economic Analysis Document," DOE/CE-0029, U.S. Department of Energy, March 1982, p. 31.

[b]The seasonal efficiency for gas furnaces is the AFUE value, and the overall efficiency for water heaters is the service efficiency as specified by the DOE test procedures.

[c]1975 rather than 1972.

[d]These values are estimates made by manufacturers in 1979.

[e]EER is the energy efficiency ratio in terms of Btu/hr of cooling output divided by watts of electrical power input. The SEER for central air conditioners is a seasonal energy efficiency ratio as specified by the DOE test procedure (see *Federal Register*, Vol. 44, p. 76700, Dec. 27, 1979).

[f]Energy factor is the corrected volume divided by daily electricity consumption, where corrected volume is the refrigerated space plus 1.63 times the freezer space for refrigerator/freezers and 1.73 times the freezer space for freezers.

Source: American Council For an Energy-Efficient Economy (1985).

SPACE HEATING

The highly efficient furnaces that first appeared on the market in 1981 incorporate forced or induced draft and the condensing of flue gases. Figure 6.1 shows the first commercial condensing furnace, which also features pulse combustion. Since this furnace was introduced, a variety of condensing furnaces has become available. Also, a prototye condensing space heater has been developed and successfully field tested (Thrasher et al. 1984). Field tests showed that condensing furnaces provide a 28% reduction in gas consumption for space heating compared with

Table 6.2. Comparison of typical and highly efficient residential products

Product	Efficiency parameter[a]	Efficiency of typical model solid in 1978	Efficiency of typical model sold in 1983	Efficiency of top models sold in 1984/85
Gas furnace	seasonal efficiency	0.64	0.70	0.94–0.97
Gas water heater	overall efficiency	0.48	0.50[b]	0.64
Electric heat pump	heating COP	1.7[b]	1.9[b]	2.6
Electric water heater	overall COP	0.81	0.83[b]	2.2[c]
Central air conditioner	SEER	7.0	8.4	15.0
Room air conditioner	EER	6.8	7.3	12.0
Top mount refrigerator/ freezer with automatic defrost	energy factor	4.8	6.5	9.8
Chest freezer with manual defrost	energy factor	11.7	11.9	22.1

[a]See Table 6.1 for definitions of the efficiency parameters.
[b]Estimates.
[c]Heat pump water heater.
Source: American Council for an **Energy-Efficient** Economy (1985).

conventional furnaces of the 1970s (Linteris 1984). This result is consistent with the laboratory test ratings. However, the durability of the sophisticated condensing furnaces, particularly with respect to the lifetime of the heat exchangers, is of some concern.

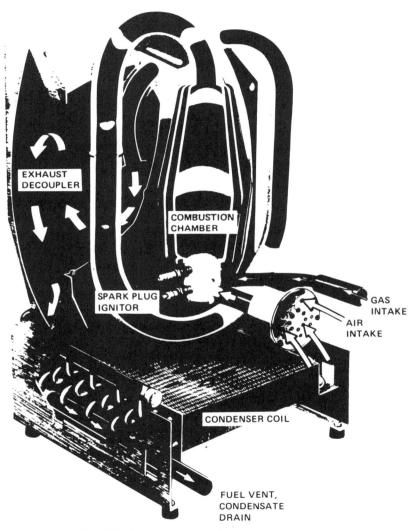

Fig. 6.1. Lennox pulse combustion gas furnace.

WATER HEATERS

The best stand-alone gas-fired water heaters on the market in 1985 feature better insulation, lower pilot rates, and improved heat transfer from the combustion gases to the water in the tank (Geller 1985). Energy efficient models have also been developed with electric ignition and a flue damper, but these features are not commonly available.

Another way to obtain high efficiencies for gas-fired water heating is to couple a hot water tank with a condensing furnace. This eliminates the substantial flue losses that occur in a conventional water heater. Such units yield an overall water heating efficiency in excess of 80% (Geller 1983). In addition, a prototype stand-alone pulse-combustion condensing gas water heater with an overall efficiency of 83% has been produced (Thrasher et al. 1984).

For electric water heating, the heat-pump water heater (HPWH) became commercially available in the early 1980s. HPWHs are now produced by a number of manufacturers as either add-on or integral units (see Fig. 6.2). They consume about 50% less electricity than conventional resistance water heaters do (Dobyns and Blatt 1984; Usibelli 1984).

The top-rated HPWH listed in Table 6.2 has a test rating significantly greater than other HPWHs. That model features a plate condenser built into the water storage tank and thicker than average foam insulation.

One concern with HPWHs is that they remove heat from the surrounding air, thereby increasing space heating requirements if located in conditioned space. Of course, they would contribute to air conditioning savings during the cooling season under the same circumstances. Simulations showed that in most locations, it is advantageous to place a HPWH outside of conditioned space (i.e., in a garage or basement; Levins 1982). To avoid "stealing" heat from interior space, a HPWH can operate with ventilation air in a house or apartment building with mechanical ventilation. Such "state-of-the-art" systems are now commonly used in new homes in Scandinavia (Geller 1985).

AIR CONDITIONERS AND HEAT PUMPS

For air conditioners and space conditioning heat pumps, large improvements in overall efficiency have been achieved through the use of larger condenser and evaporator coils; more-efficient motors; oversized, underrated, and rotary-type compressors; and improved controls (Geller 1985). In addition, some highly efficient models use two-speed compressors, dual compressors, or continuous-speed modulation (Lennox 1984), providing a much better matching of air conditioner or heat pump output to the load and thereby reducing cycling losses (Fig. 6.3).

Energy efficient air conditioners may not provide adequate dehumidification, particularly in homes of high thermal integrity in hot, humid climates, where the ratio of latent to sensible cooling loads is high (Florida Solar Energy Center 1983; Fairey 1984). Efforts to overcome this problem include the introduction of units with special humidity control features (Lennox 1984), the development of a heat pipe assisted air conditioner with high dehumidification capability (Khattar 1985), and the development of desiccant-assisted systems (Florida Solar Energy Center 1983).

Condensing Furnaces

In an ordinary domestic furnace, about 25% of the fuel's heat content is carried up the chimney by the hot exhaust gases. Research was carried out in the sixties and seventies to develop furnaces in which most of this heat would be recaptured. The trick is to cool the exhaust gases and condense the water vapor with high heat transfer, but that cooling requires special combustion conditions and heat exchangers. Much progress was made at the laboratory level including the development of a process known as pulse combustion.

In the late seventies the gas industry (through GRI and AGA) solicited the assistance of a major gas-furnace manufacturer to move the pulse combustion condensing furnace to commercialization. Final technical obstacles, such as size and noise, were overcome, and prototype units were fabricated for field testing in 1979.

Lennox, the company involved in developing the pulse-combustion furnace, began commercial production in Spring 1982. Other manufacturers began producing condensing gas furnaces in late 1982 and in 1983. All of these units have rated efficiencies in excess of 90%, compared to 63% for typical domestic furnaces sold in the 1970s. By early 1984, the fraction of the gas-furnace market captured by condensing units had already risen to 10%.

The performance of conventional heat pumps in the U.S. should continue to improve through ongoing R&D to reduce dynamic energy losses, develop units with capacity modulation and improved controls, and develop new refrigerant mixtures (Lannus 1985; Fairchild 1985). In addition, R&D on novel refrigeration cycles could lead to further improvements.

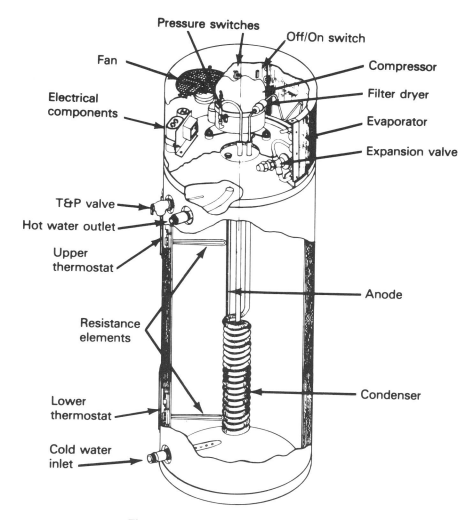

Fig. 6.2. Electric heat pump water heater.

In Japan, manufacturers are starting to include solid-state inverters in their air conditioner and heat pump systems. This provides capacity modulation through continuous speed variation. The Japanese are also producing systems in which the refrigerant is circulated to independently controlled indoor heat exchange coils, providing zonal space conditioning within the house (Geller et al. 1986).

Evaporative coolers have long been used for cooling in hot, dry climates. Evaporative coolers are efficient, consuming 80–90% less electricity than conventional vapor compression AC systems (Geller et al. 1986). But conventional, direct evaporative coolers add considerable moisture to indoor air and this has limited their popularity in the U.S. Indirect, two-stage evaporative coolers overcome the moisture problem. Indirect evaporative coolers are now marketed for commercial buildings; systems for residences are expected on the market soon (Geller et al. 1986).

REFRIGERATORS AND FREEZERS

The energy consumption of refrigerators and freezers declined significantly in the past decade. These improvements occurred through the use of better insulation and more-efficient motors and compressors (Geller 1985).

Figure 6.4 shows the progress that has been made for top-mount refrigerator-freezers with automatic defrost. In 1985, the most efficient top-mount refrigerator-freezer with automatic defrost consumed 750 kWh/yr. This value compares with 1990 kWh/yr for the typical top-mount model made in 1972 and 1150 kWh/yr for the typical model made in 1983 (Association of Home Appliance Manufacturers 1984).

A prototype 18-ft^3 two-door model has an electricity consumption rating of 620 kWh/yr. This unit has a highly efficient motor-compressor and separate cooling coils in the refrigerator and freezer boxes (Geller 1985). The improved compressor has not found its way into commercial models so far, although the advanced refrigeration system has been used commercially.

Well-designed, extremely efficient refrigerators are being custom-built by a small company in California primarily for use with photovoltaic power systems. These units feature high levels of insulation, separate cooling systems for the refrigerator and freezer boxes, and better component placement (e.g., the motor-compressor is above rather than below the food storage space). A 16 ft^3 two-door model consumes only 240 kWh/yr (Geller 1985).

Japanese refrigerators in some cases are more efficient than their American counterparts (Goldstein 1984), and much progress has been made towards the development of highly efficient refrigerators in Denmark (Norgard and Heeboll 1983). Promising research has also been carried out in the U.S. in such areas as new refrigerant mixtures (Levins 1984). Unfortunately, there is very little information publicly available on the performance of refrigerators from around the world tested under standardized conditions.

COOKING EQUIPMENT

For gas cooking, most stoves now include electric ignition. This lowers overall gas consumption by 35 to 45% compared to older gas stoves with standing pilots (Geller 1985). An advanced gas burner known as an infrared jet-impingement burner has been developed in recent years. This burner is up to 50% more efficient than conventional gas burners and has much lower NO_x emissions (Gas Research Institute 1984). Field tests of stoves equipped with this burner began in 1984.

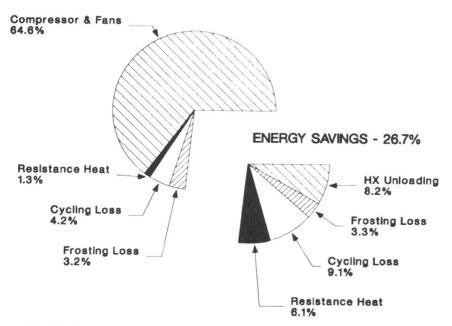

Fig. 6.3. Electricity use and savings for a continuously modulating heat pump.

For electric cooking, microwave ovens are considerably more efficient than conventional electric ovens. Cooking tests with actual food items show that use of a microwave oven reduces electricity consumption by 25 to 50% relative to cooking only with a conventional electric range (Ludvigson and VanValkenburg 1978).

The induction cooktop is another advanced technology now commercially available. An induction cooktop features magnetic coils that induce currents in iron or steel pans, eliminating losses from the heating elements. Induction cooking should lead to a 15–25% electricity savings for stovetop cooking (Geller et al. 1986).

OTHER APPLIANCES

The majority of the energy consumed by clothes washers and dishwashers is in the form of energy to heat the water they use. Hot water consumption can be lowered by using cold water for clothes washing and/or rinsing, front-load clothes washers, and dishwashers with shorter or fewer subcycles.

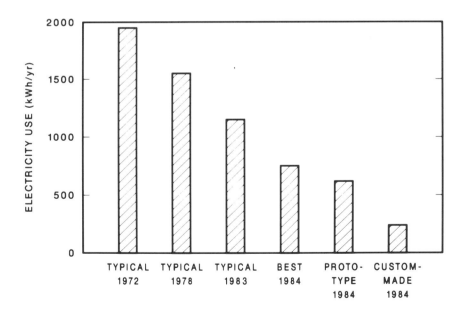

Fig. 6.4. Progress in reducing electricity use for top-mount freezer, automatic defrost refrigerators sold in the U.S. (Geller 1984).

Several steps have been taken to improve the efficiency of clothes dryers. Standing pilots have been eliminated. Dryers with automatic termination controls are available that reduce energy consumption by 10 to 15% (Levins 1980). In addition, prototype electric clothes dryers have been developed that use either microwave radiation or a heat pump for drying. In both cases, electricity consumption is estimated to be 50 to 75% lower than that of a conventional electric dryer (Geller 1985).

LIGHTING

Incandescent bulbs typically provide only 15 lumens per watt, with about 90% of the power drawn by the bulb given off as heat. Considerable attention has been devoted to increasing the efficiency of residential lighting (Verderber and Rubinstein 1983).

Compact fluorescent lamps are now produced by major lighting manufacturers in Europe, the U.S. and Japan. They are available either as a complete unit containing tube and ballast or as a separate conversion base and plug-in tube. One advanced model manufactured in the U.S. by Philips includes a high frequency electronic ballast. The electronic ballast reduces size, weight, and power consumption.

Compact fluorescent lamps typically consume 60–75% less power than incandescents for the same light output (Geller et al. 1986). Unlike conventional fluorescent lights, compact fluorescents provide "warm" light similar to incandescents. In addition, they last 5–10 times longer than incandescent bulbs.

A drawback to today's compact fluorescents is that most models are larger than incandescent bulbs. However, a West German company has developed a compact fluorescent bulb of equivalent size to incandescents. Also, the first cost for a compact fluorescent lamp is $10–20. But, as shown in Table 6.3, the bulbs are economical on the basis of life-cycle cost.

Incandescent bulbs are being improved. A spherical bulb with a heat reflecting inner coating has been developed. It uses half as much electricity as ordinary incandescents for the same light output (Duro-Test Corp. 1982). This bulb is expected to be marketed in the U.S. in 1986.

RETROFIT DEVICES

Many retrofit devices and techniques are available for improving the efficiency of residential equipment. Table 6.4 lists the more popular retrofit devices for single-family residences and the energy savings obtained. In most cases, the savings values are based on field monitoring experiments in substantial numbers of homes. The bulk of the retrofit devices are for space heating systems, where the savings potential can be as great as 24%. The retrofit devices have been well studied, and their advantages, disadvantages, and limitations have been identified (Kweller and Silberstein 1984).

Table 6.3. Cost effectiveness of some highly efficient appliances
and light bulbs in the U.S.[a]

Model	Increased first cost[b] (1985 $)	Annual electr. savings (kWh/yr)	Simple payback period (yrs)	Internal rate of return (%/yr)
Whirlpool ET17HK1M refrigerator/freezer	60	330	2.4	45
DEC International heat pump water heater	1200	2500[c]	6.3	15
Lennox HS-14 Power Saver central air conditioner	1000	2100[c]	6.3	14
Friedrich SM10G10 room air conditioner	130	210[c]	8.1	12
Philips SL18 compact fluorescent lamp[d]	18	57	4.0	24
Panasonic 17 W fluorescent lamp[e]	10	43	3.0	34

[a]Based on a first year electricity price of $0.076/kWh and a 2%/yr real electricity price escalation rate. The first year energy price was the national average in 1984 and the escalation rate is based on utility industry forecasts.
[b]The cost difference is relative to a model of standard efficiency made by the same manufacturer. Cost data were obtained from dealers and contractors in the Washington, D.C., area.
[c]For air conditioning and hot water heating, the estimated use in a typical U.S. household is assumed.
[d]It is assumed that this 18 watt fluorescent bulb replaces a 75 watt long-life incandescent bulb and is used for about 2.7 hours/day.
[e]It is assumed that this 17 watt bulb replaces a 60 watt incandescent bulb and is used 2.7 hours/day.
Source: American Council for an Energy-Efficient Economy.

Research and demonstration has shown that large energy savings can often be achieved with retrofits to the space-conditioning and water heating systems in multifamily housing. For example, the retrofit of "outdoor reset" controls, which vary the water temperature in hydronic heating systems depending on the ambient temperature, reduce fuel consumption by 10 to 20% (Hewett and Peterson 1984). In another example, modifications to the boiler, distribution system, and steam radiators in an apartment building in New Jersey provided a 50% reduction in heating fuel consumption (DeCicco et al. 1984). Nonetheless, it appears that information is still lacking on the broad potential for

Table 6.4. **Energy savings from furnace, boiler, air conditioner, and electric water heater retrofit devices**

Retrofit measure	Applicable products[a]	Applicable fuel[a]	Estimated energy savings[b]	
			Range (%)	Average (%)
Derate	F,B	O,G	0–19	9
Electro-mechanical vent or flue damper	F,B	O,G	0–14	6
Derate plus E/M damper	F,B	O,G	8–20	
Thermal damper	F,B	O,G	3–8	
Intermittent ignition device	F,B	G	2–6	
Flue gas heat reclaimer or economizer	F,B	O,G	9–20	8
Flue gas heat reclaimer with derate and damper	F,B	O,G	21–24	
High-speed flame retention burner	F,B	O	5–22	14
Insulation wrap (R-11)	WH	E,G	5–12	
"Heat trap" convection inhibitor in hot water line	WH	E,G,O		2
Heat recovery unit	WH and CAC/HP	E	25-60	

[a]F—furnace, B—boiler, WH—water heater, CAC/HP—central air conditioner or heat pump, O—oil, G—gas, and E—electric.
[b]The energy savings are relative to standard models typically purchased in 1980.
Source: Geller (1983).

equipment retrofits in multifamily buildings, interactions between different measures, and energy savings over an extended time (Harris and Blumstein 1984).

C. TEST PROCEDURES AND FIELD PERFORMANCE

The limited number of studies comparing appliance energy consumption in the field to test ratings yielded varying results. Some studies of the electricity consumption of refrigerators found actual consumption to be greater than the ratings (Lawrence 1982; Topping 1982). In other cases, actual consumption was lower than the test ratings (Middleton and Sauber 1983). More-efficient models tend to have a lower actual energy consumption relative to the standardized ratings than do models of average efficiency (Geller 1985).

Clearly, users can have a significant impact on appliance energy consumption. Field measurements with identical refrigerator models in different households in Florida showed electricity usage varying by a factor of two (Lawrence 1982). In another experiment, different cooks prepared the same meal on identical electric ranges, and energy consumption varied by as much as 50% (Fechter et al. 1979). An evaluation of the field performance of refrigerators in ten households in New Jersey showed that the temperature difference between the kitchen and the refrigerator has the most effect on energy consumption, while the number of door openings has a lesser but still important effect (Chang and Grot 1979).

Less is known about how specific operating factors affect the actual energy consumption of appliances and space-conditioning equipment (Harris and Blumstein 1984; Clear and Goldstein 1980; Usibelli 1984). Unfortunately, many field tests are carried out without a detailed description of the usage conditions. Also, little information is available on performance degradation and changes in appliance energy consumption over time.

Increasing appliance efficiency can affect space-conditioning requirements. If highly efficient appliances do not dump as much waste heat into the interiors of buildings any more, space heating requirements could increase. But building simulation studies (Palmiter and Kennedy 1983; Corum 1984) show that even if internal thermal gains are reduced through the use of more-efficient appliances, substantial overall energy savings still result because some thermal gains occur in unconditioned space and because space heating is required only part of the year. Of course, the overall benefit from the use of more-efficient appliances is site specific and increases where space cooling is more important. However, the interaction between appliance efficiency and space conditioning has not yet been addressed in field studies.

D. ECONOMICS OF ENERGY-EFFICIENT APPLIANCES

DOE sponsored assessments of the additional manufacturing and retail cost for highly efficient residential appliances and HVAC equipment (A. D. Little 1982; Department of Energy 1983). These studies show that efficient appliances cost more than models of average efficiency.

Limited surveys have been conducted to compare the first costs of energy efficient and standard models. Table 6.3 shows the results of one limited survey in the Washington, D.C., area to establish the extra first cost for highly efficient products on the market in 1984 (Geller 1984). The cost increase varies from product to product. Compact fluorescent lamps cost at least 10 times as much as ordinary incandescent bulbs. The top-rated heat-pump water heater has a first cost about four times that of an ordinary resistance water heater. For the highly efficient central air conditioner, the first cost is 50 to 70% greater than that of standard models. On the other hand, the highly efficient refrigerators and freezers are only about 10% more expensive than their counterparts of average efficiency.

The extra costs shown in Table 6.3 may not necessarily reflect the increased production cost, but rather what the manufacturers feel the more efficient products can command in the marketplace. They may also reflect a desire to rapidly recoup R&D costs or demonstrate the effects of the lack of economies of scale during initial production. The comparison of retail prices is complicated by differences in features between models; more work is needed on the cost-efficiency relationships for residential appliances and HVAC equipment.

Nevertheless, highly efficient appliances are cost-effective when life-cycle cost is considered (Geller 1983; Meier et al. 1983; Science Applications, Inc. 1982). Table 6.3 shows the cost effectiveness of buying highly efficient models. The simple payback period on the additional first cost ranges from 2 to 8 years. On the average, the payback period is less than half the assumed product lifetime. The rate of return on the extra first cost, which is above inflation and tax-free, ranges from 12%/year to 45%/year.

Investing in more-efficient appliances should not be subject to the same uncertainties that plague some conservation opportunities (e.g., residential retrofits). In particular, obtaining the expected energy savings is very probable when a highly efficient furnace, water heater, or refrigerator is purchased. However, this expectation needs to be confirmed in field studies.

E. INSTITUTIONAL BARRIERS

Despite the fact that more-efficient appliances are very cost-effective for consumers, assessments of overall market behavior in the purchase of appliances show similar results as obtained in some studies of investments in the thermal integrity of homes: a major underinvestment in energy efficiency (Science Applications, Inc. 1982; Meier and Whittier 1982). These studies show that for most products, implicit discount rates are more than 50%/year. Moreover, implicit discount rates for appliances by and large did not decline during the 1970s (Ruderman et al. 1984).

Several factors contribute to this market failure (Geller 1983; Stern and Aronson 1984; California Energy Commission 1983). Consumers generally have a poor understanding of the opportunties to save energy and money through investing in more-efficient equipment. Even if consumers are concerned about their energy bills, they do not have access to information regarding the energy consumption and operating costs of individual appliances (Stern and Aronson 1984).

Many decisions regarding the selection of appliances and space conditioning systems are made by third parties: builders, contractors, and landlords. For example, a survey of 950 households conducted in 1979 indicated that fewer than 25% of all central heating systems are purchased directly by the user (Reid 1981). Substantial evidence indicates that third-party purchasers tend to buy low-first-cost, less-efficient equipment (Geller 1983; California Energy Commission 1983). This is logical because third-party purchasers are first-cost sensitive and do not pay operating costs.

The nature of purchase decisions also limits consideration of efficiency in residential appliance decisions. Consumers looking for a replacement furnace, water heater, or refrigerator are often concerned with restoring service as soon as possible. Shopping is done in haste, and consumers may accept whatever the salesperson can deliver immediately. Even when time is available to consider operating costs, other factors (e.g., design features, appearance, and durability) play a larger role in purchase decisions (Stern and Aronson 1984; Good Housekeeping Institute 1982).

Other barriers that inhibit the construction of more-efficient homes also apply to the purchase of appliances and HVAC equipment. Table 6.3 shows that highly efficient furnaces, air conditioners, and water heaters can have an extra first cost of around $1000. Lack of available capital and uncertainties regarding the recovery of this extra first cost at the time of home sale might prevent some consumers from purchasing highly efficient models.

F. UNRESOLVED ISSUES

Many of the unresolved issues concerning energy efficient residential appliances and space conditioning equipment have been touched upon already.

One set of issues relates to test procedures and field performance. The rating procedures need to be reviewed in light of efficiency improvements and changing usage conditions during the past decade. Further comparisons of the test ratings and actual field performance are needed. Test procedures may need to be revised so they more accurately predict actual performance. Also, further study of the interaction between improving appliance efficiency and space conditioning requirements is needed and should be incorporated into economic and program assessments.

Although remarkable progress has been made in developing and commercializing highly efficient residential products in recent years, much more can be done technically. Furthermore, economic analysis shows that few of the currently available top-rated products are at the point where further efficiency increases are not cost-effective. Product development should continue at least in the following areas:

- Refrigerators and freezers with more-efficient components, better insulation, and improved refrigeration-system design

- Air conditioners with high dehumidification capacity

- Fuel-fired water heaters with flue-gas condensation

- Heat-activated (fossil-fuel-fired) heat pumps

- Advanced clothes dryers that use microwave radiation or a vapor-compression cycle

- Clothes washers and dishwashers that use less hot water

- Integrated appliances (such as combined water heating and refrigeration systems or mechanical ventilation equipment with heat recovery)

- Improved controls for residential appliances.

Along with the promising near-term, product-oriented appliance projects, fundamental research areas exist that could lead to performance improvements over the long term. Included in this category are advanced refrigerant mixtures and new insulation materials.

Another set of issues concerns the maintenance requirements and lifetime of such new technologies as condensing furnaces and heat-pump water heaters. Also, like building retrofit measures, little is known about the durability and long-term performance of appliance retrofit devices.

Finally, given the growing interest in appliance efficiency and the increasing international trade in appliances, direct comparisons of state-of-the-art technologies around the world are needed. Unfortunately, efficiency test procedures vary throughout the world; hence little comparable performance data are available for American and foreign appliances at the present time.

7

NEW AND EXISTING COMMERCIAL BUILDINGS

A. DESCRIPTION AND CURRENT PATTERNS OF ENERGY USE

The term "commercial buildings" covers a wide range of nonresidential building types. The Energy Information Administration (EIA) conducted a nationwide survey regarding energy use and other aspects of nearly 5600 commercial buildings as of 1979 (Energy Information Administration 1983a). Table 7.1 shows the estimated breakdown of the *total* floor space in the commercial sector by building type based on this survey.

Office buildings represent the largest fraction of total floor area (17%), followed by retail/services buildings (16%). Warehouse and storage, educational, and assembly buildings each account for more than 10% of the total commercial sector floor area as well. The estimated average floor area per building (11,900 ft^2) is three times as great as the median because of a small number of very large buildings.

Figure 7.1 presents estimates of average end-use consumption by building type. The estimates are based on 1979 billing data collected as part of the EIA survey. The average consumption for all commercial buildings is 115,000 Btu/ft^2 of floor area. Health care facilities (hospitals and clinics) are more than twice as energy intensive as the commercial sector in general. However, total energy use in most of the subsectors is within 25% of the overall average. Energy consumption per square foot is slightly higher than the national average in the Northeast (124,000 Btu/ft^2) and North Central (130,000 Btu/ft^2) regions, and below average in the South (106,000 Btu/ft^2) and West (86,000 Btu/ft^2) (Energy Information Administration 1983b). (EIA collected energy consumption and other data from the same set of buildings again in 1983, but the results of this second survey were not available soon enough to include here.)

Figure 7.2 and Table 7.2 present information on energy consumption by building size. Average consumption per square foot generally declines as building size increases. Although smaller buildings (less than 10,000 ft^2) predominate in number, they represent only 21% of total floor area and 27% of total energy consumption.

Just as total energy consumption can vary substantially among building types, so too can end use. Table 7.3 shows one estimate of how energy is used in new commercial buildings of different types. In this breakdown, the energy required to operate motors, fans, etc. in the HVAC system is separate from heating and cooling end uses. The fraction of energy required for heating or cooling varies by a factor of four among building types. Lighting in almost all commercial categories takes a significant share of the total energy demand. Heating and cooling loads are low where people are not routinely present (storage buildings) or where people tend to stay for only a specified or scheduled time (assembly buildings), and they are highest for hotels and motels.

The values given in the preceeding tables and figures are averages. Energy consumption may be higher or lower for a specific building depending on location, building characteristics, occupant behavior, and other factors (Fig. 7.3).

Table 7.1. Estimated breakdown of commercial buildings, 1979

Building type	Total buildings (10^3)	Average area per building (10^{3}ft^2)	Total area (10^{9}ft^2)	(%)
Office	600	13.6	8.2	17.2
Retail/services	714	10.7	7.7	16.1
Warehouse and storage	430	14.1	6.1	12.8
Education	161	36.2	5.9	12.4
Assembly	448	11.2	5.0	10.5
Residential	347	9.0	3.1	6.5
Other	237	13.2	3.1	6.5
Lodging	101	19.9	2.0	4.2
Food sales	366	5.1	1.9	4.0
Automotive sales and service	401	4.5	1.8	3.8
Health care	44	38.5	1.7	3.6
Vacant	146	8.7	1.3	2.7
Total	3,995	11.9	47.7	

Source: Energy Information Administration (1983a).

B. SAVINGS POTENTIAL IN NEW BUILDINGS

OVERALL BEPS RESULTS

The BEPS program (described in Chapter 9) also studied the energy performance of new commercial buildings (*Progressive Architecture* 1982a and b, and 1983a and b; Stoops et al. 1984). The central questions were how much energy do new buildings demand and how much of this demand can be reduced through energy-conscious design.

The first phase of the BEPS research effort focused on development of "baseline" data about the energy performance of recently designed buildings. Initially, 1600 buildings designed and built in the mid-1970s were identified, and the design energy performance of each was determined with a simulation model.

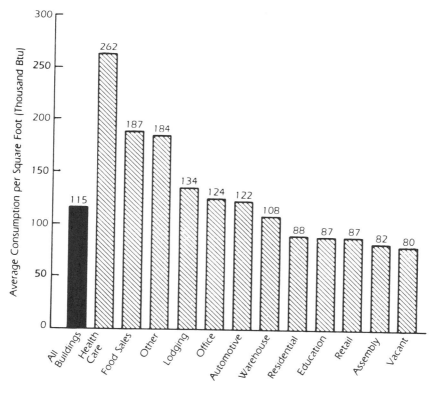

Fig. 7.1. Energy use per unit floor area by commercial-building type, 1979 (Energy Information Administration 1983b).

Table 7.4 shows the average site energy use estimated for the original designs. These values are averaged over both the range of buildings and the various locations considered. Except for hospitals, total annual energy use generally ranges from 60 to 100 thousand Btu/ft^2. The energy performance of these new designs is generally superior to that in the overall commercial building stock. (Compare Fig. 7.1 and Table 7.4.)

The BEPS effort then compared the energy performance of the original buildings with that of buildings meeting the minimum ASHRAE

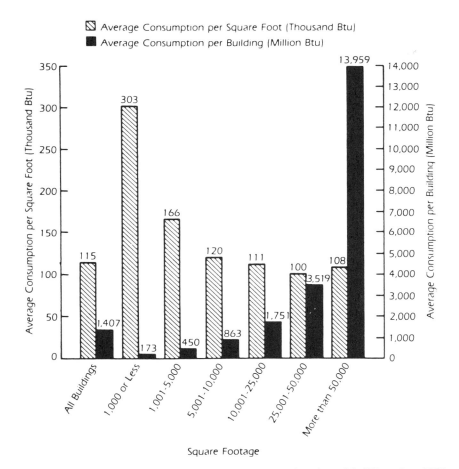

Fig. 7.2. Energy use in commercial buildings as a function of building size, 1979 (Energy Information Administration 1983b).

Table 7.2. Prevalence of buildings and energy consumption as a function of building size, 1979

Size (ft^2)	Percentage of total buildings	Percentage of total floor area	Percentage of total consumption
1,000 or less	15	1	2
1,001–5,000	42	9	13
5,001–10,000	19	11	12
10,001–25,000	14	18	18
25,001–50,000	5	15	13
More than 50,000	4	45	42

Source: Energy Information Administration (1983b).

Table 7.3. End-use energy consumption for new commercial buildings

Building type	End-use energy consumption as a fraction of total energy (%)					
	Heating	Cooling	HVAC[a]	Lights	DHW[b]	Other
Assembly buildings	32	10	14	32	9	3
Elementary schools	41	16	9	24	9	0
Secondary schools	45	17	7	23	8	1
Hospitals	41	18	12	18	2	10
Clinics	29	18	9	28	6	11
Stores	31	13	10	36	5	6
Shopping centers	28	14	8	41	7	2
Small offices	27	20	6	36	11	0
Large offices	24	17	12	35	4	9
Storage buildings	64	7	5	22	1	1
Hotels/motels	15	29	16	18	8	14
Nursing homes	38	14	10	23	4	12

[a]Energy to drive pumps, fans, etc in the HVAC system is included separately from heating and cooling.

[b]DHW is domestic hot water.

Source: American Institute of Architects (1982).

standards (Standard 90-75) for commercial buildings. Table 7.4 shows that designs meeting the ASHRAE standards would use as much as 40% less energy than the original designs. If the results are averaged over the different building types, meeting the ASHRAE standards would lead to a 22% reduction in energy use.

In a second phase of the study, a subset of 168 buildings was statistically selected, and the original architects and engineers participated in a workshop on energy efficient design concepts and strategies. Those design teams then redesigned the buildings while receiving peer review and feedback from experts. The redesigns generally had to adhere to the building owner's functional requirements, site restrictions, and construction budget.

Table 7.4 gives the estimated energy performance of the redesigned buildings. Averaged over the different building types, the new designs used 41% less energy than the original designs, showing that significant reductions in energy consumption are possible with existing design practices. In addition, these reductions were achieved within the present structure, organization, and practice of the building industry and were accomplished with little increase in the first cost. In some cases the first cost even decreased.

An examination of the strategies employed in the BEPS redesign indicates how it is possible to reduce the energy consumption of new commercial buildings so dramatically with a modest effort. The case studies presented below for three building types were taken from a series

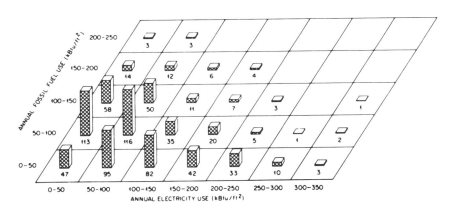

Fig. 7.3. Relationship between fossil-fuel use and electricity use in school buildings in ten states (Hirst, Carney and Knight 1981). The numbers and heights of each box refer to the number of buildings in that cell.

of articles published by *Progressive Architecture*. They illustrate the variety of options that are available for improving the efficiency of commercial buildings and the importance of matching design features to the particular needs and uses of a building.

Office Buildings

Twenty-two buildings, ranging from a 3,800-ft^2 1-story branch bank to a 637,000-ft^2 29-story office tower, were redesigned, and the energy-efficiency improvements were analyzed (*Progressive Architecture* 1982b). Half of the sample were buildings with 50,000 ft^2 or less located in a wide range of climates. Redesign produced energy savings in excess of 40% with an estimated 3.5% increase in cost of construction.

Office buildings are characterized by daytime occupancy and high internal loads. If the envelope is reasonably tight, internal heat gains can lower the balance point (the outside temperature below which a building needs to be mechanically heated) to 40°F or below. Control of heat gain becomes a critical design strategy for such buildings. At night during the

Table 7.4. Comparisons of commercial energy consumption in the BEPS research project[a]

Building type	Original design kBtu/ft^2	ASHRAE 90-75		Redesign	
		kBtu/ ft^2	Percentage change from original	kBtu/ ft^2	Percentage change from original
Small office	77	58	25	39	49
Large office	69	48	30	40	42
Stores	87	67	23	58	33
Shopping centers	98	80	18	65	34
Hotel/motel	96	74	23	68	29
Elementary schools	58	52	10	31	47
Secondary schools	85	60	29	51	40
Warehouses	62	37	40	29	53
Assembly buildings	86	75	13	48	44
Clinics	86	74	14	51	41
Nursing homes	92	58	37	55	40
Hospitals	245	250	(-2)	142	42

[a]Energy demand levels are estimates of end-use energy consumption.
Source: Progressive Architecture (1984); Stoops et al. (1984).

heating season, heat gains are normally absent, and conduction and infiltration heat losses become important. In warm climates, control of solar gains tends to be the dominant energy concern.

In office buildings, high turnover in occupancy and the need for moveable partitions lead to grid lighting systems controlled from a central bank of circuit breakers. With control removed from the occupants, lights are more likely to be left on when space is unoccupied, and lighting levels can be far in excess of what is needed. Lighting, of course, contributes to the internal heat gain and thus to cooling loads in many situations.

Occupancy in large office buildings has another dimension; many such buildings include not just office space, but restaurants, recreation areas, assembly halls, and the like. This variety in occupancy creates different energy design problems and opportunities.

The energy improvement strategies used in the BEPS redesigns of office buildings included a shift towards more compact forms, optimizing thermal gains and losses, beneficial orientation, and daylighting.

Envelope redesign focused on reorienting the entire building, reducing or reorienting the glazing, employing different types of glazings, and using interior shading devices. More than half of the office redesigns reoriented their buildings in an elongated form along the east-west axis for solar control. A small office building in Wisconsin used light scoops and a sunspace facing due south for maximum solar contribution. A large office building in Connecticut was turned towards the sun, its windows were recessed, and its form was elongated. In Bakersfield, California, where cooling loads dominate, the surface-to-volume ratio of a small office building was reduced.

In nearly all of the new designs, average lighting capacity was reduced from 2.7 to 1.7 W/ft^2. The various strategies for redesigning lighting included reduction of illumination levels, more-efficient light fixtures and controls, task lighting, reduction in the use of incandescent lights, and daylighting.

HVAC systems were changed in many of the buildings. Variable-air-volume (VAV) systems were favored in large offices, while heat pumps were favored in small offices. Five of the eleven small offices switched to the use of hydronic heat pumps. More than half of the original office building designs used an economizer cycle for cooling with outside air when possible. This strategy was also adopted by the other buildings in the redesign.

School Buildings

Eleven elementary schools and eleven secondary schools were redesigned (*Progressive Architecture* 1983b). Schools have intermittent and irregular usage patterns, high ventilation rates, and high lighting levels.

Several redesign strategies were used to achieve the 47% energy improvement in elementary schools and the 40% improvement in secondary schools. Control of heating and cooling loads was achieved in warm climates through greater R values for walls. Walls in buildings in cold climates were already reasonably well insulated. Thermal losses through windows were reduced by upgrading from single to double glazing.

Passive direct-gain systems were employed where southern exposure was available. The inherent thermal mass in the buildings, in combination with solar energy, provided for the unoccupied heating loads.

Lighting energy use was reduced by lowering lighting levels to 1.5 W/ft^2 in conjunction with extensive use of daylighting. The types of daylighting employed were side lighting, clerestories, monitors, and skylights.

HVAC strategies focused on operation and control modifications, control of outside air, and use of heat recovery. The original HVAC systems remained unchanged, and potentially effective strategies, such as modular boilers or plant equipment staging, were not used. Changing the night heating setpoint to 60°F or less and the cooling setpoint to 78°F or more was frequently employed.

Hospitals and Clinics

Hospitals are by far the most energy-intensive type of commercial building. The 24-hour-a-day occupancy and strict environmental, life-safety, and regulatory requirements heavily affect design requirements and strategies for achieving greater energy efficiency. The BEPS redesign analysis included 10 clinics and 11 hospitals (*Progressive Architecture* 1983a). The overall 42% savings was achieved primarily through reductions in heating energy, related auxiliary equipment (such as fans and pumps), and lighting.

Hospital light levels were reduced in support areas, such as corridors and nurses' stations. Improved lighting controls, such as time clocks and local two-step switching, were also employed. In office and administrative areas, VAV systems were employed, and task lighting was used where applicable. In many redesigns, patient rooms were located around an interior courtyard or atrium. This arrangement made the overall building form more compact and provided natural lighting. Functional spaces were assigned to a particular type of HVAC system on the basis of operating hours, temperature setpoints, and humidification requirements.

Clinics are similar to office buildings from an energy performance point of view with the exception of nighttime and weekend occupancy. The redesign strategies included optimization of gains and losses through compact building forms, improvement of U values in walls and glazing, use of fluorescent fixtures with energy efficient ballasts, use of exterior shading devices, inclusion of moveable insulation on the windows, and reduction of glazing.

Nearly all the clinics in the original designs used constant-volume or dual-duct HVAC systems. In the redesign, six of the projects used VAV systems, while three used heat pumps. Widespread use of economizer cycles, changes in heating and cooling setpoints, night setback strategies, and reduction of outside air quantities contributed to the improvements in energy efficiency.

ACTUAL PERFORMANCE OF NEW BUILDINGS

Several recently constructed buildings for which actual energy consumption data are available show that the performance of the BEPS redesigns can be improved upon. One set of buildings is associated with the U.S. Department of Energy's Passive Solar Commercial Building Program, which attempted to design, construct, and monitor 23 buildings with low conventional energy inputs (Department of Energy 1983). The projects covered a range of building types, sizes, and locations (Table 7.5). Nineteen of the projects were new, and four involved major retrofits. Unfortunately, few larger buildings were included in the program.

The buildings incorporated passive solar features and traditional energy efficiency measures, such as high levels of insulation, heat pumps, lighting, and HVAC controls. The buildings were designed with expert technical assistance on energy related issues provided to the design teams.

As of early 1985, 19 of the buildings had been constructed, and performance monitoring had been completed or was underway. Figures 7.4 and 7.5 illustrate two of the state-of-the-art buildings.

Table 7.6 includes data on the first cost of the new buildings and on the predicted and actual energy consumption, where available. Actual energy use is based on a full year of monitored consumption while the buildings were occupied. For the 13 buildings listed in Table 7.6, construction cost averaged $65/ft^2. A comparison with the construction cost for ordinary buildings of similar type, size, and location showed that

Table 7.5. Buildings designed as part of the Passive Solar Demonstration Program

Type building	Floor area (ft²)	Location
Schools and libraries		
Abrams Primary School	27,400	Ala.
Blake Avenue College Center	31,000	Colo.
Community United Methodist Church	5,500	Mo.
Mt. Airy Public Library	13,500	N.C.
Princeton School of Architecture	13,700	N.J.
Southwest Woodbridge School	43,000	Calif.
St. Mary's School Addition	8,800	Va.
Two Rivers School	15,750	Ark.
Office buildings		
Johnson Controls Branch Office	15,000	Utah
Princeton Professional Park	64,000	N.J.
Retail		
Kieffer Store Addition	3,200	Wis.
Wells Security State Bank	5,500	Minn.
Community centers/health care		
Deadwood Creek Community Center	3,500	Oreg.
RPI Visitor Information Center	4,300	N.Y.
Shelly Ridge Girl Scout Center	5,700	Penn.
Willow Park II Community Center	4,100	Tex.
Comal County Mental Health Center	4,800	Tex.
Essex Dorsey Senior Center	13,000	Md.
Airport terminals		
Gunnison County Airport	9,700	Colo.
Walker Field Terminal Building	66,700	Colo.
Other		
Philadelphia Municipal Auto Shop	57,000	Penn.
Touliatos Greenhouse	6,000	Tenn.

Source: Department of Energy (1983).

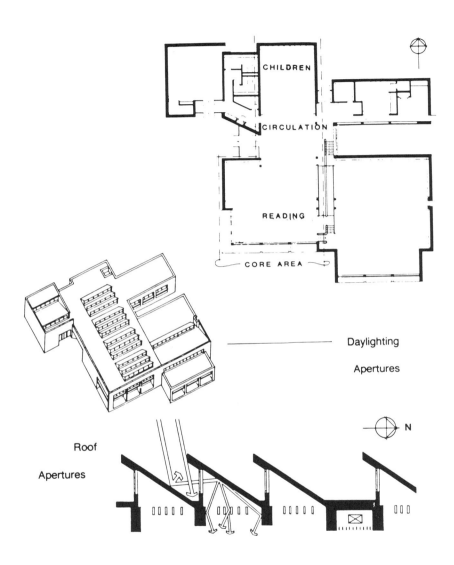

CHILDREN

CIRCULATION

READING

CORE AREA

Daylighting

Apertures

Roof

Apertures

N

Fig. 7.4. Example of a passive-solar library.

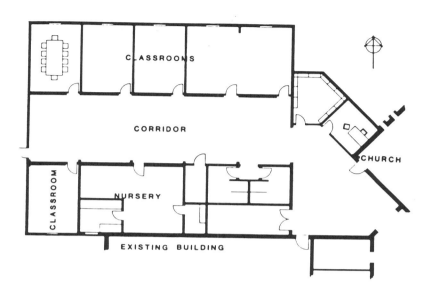

Fig. 7.5. Example of a passive-solar church.

most of the passive solar buildings do not cost more to build than conventional buildings (Gordon et al. 1985).

The predicted energy use of the new buildings was only 32 kBtu/ft2 on average, just 28% of the average consumption of ordinary commercial buildings. For the buildings where monitored data is available, consumption averaged 41 kBtu/ft^2. This value is greater than predicted but is still substantially below the energy consumption of conventional commercial buildings. The discrepancy between the predicted and actual values is attributed to unanticipated building use and to weaknesses in the predictive tools (Gordon et al. 1985).

Performance monitoring showed that heating, cooling, and lighting energy consumption were all reduced by approximately 50% relative to ordinary buildings. Because of careful building design, the heavy reliance on daylighting and passive solar heating was *not* accompanied by an increase in cooling requirements. However, energy consumption in the "other" category (for operating fans, pumps, appliances, office equipment, etc.) did increase in the passive solar buildings.

Table 7.6. Cost and performance of new commercial buildings constructed as part of the DOE Passive Solar Program[a]

Project	Cost ($/ft^2)	Predicted energy use (kBtu/ft^2)	Actual energy use (kBtu/ft^2)
Johnson Controls Office	57	51.0	35.7
Community United Methodist Church	47	16.0	20.2
Mt. Airy Library	88	17.4	26.0
Gunnison Airport	80	66.7	70.6
RPI Visitor Center	81	28.5	52.0
Wells Security State Bank	64	25.6	55.9
St. Mary's Gymnasium	74	27.1	27.3
Essex Dorsey Senior Center	65	40.3	
Shelly Ridge Girl Scout Center	85	35.0	
Abrams Primary School	36	22.2	
Blake Ave. College Center	59	33.0	
Princeton Professional Park	46	15.0	
Walker Field Terminal	60	42.0	

[a]Energy use values correspond to end-use energy consumption.
Source: Gordon et al. (1985).

Some of the principal conclusions drawn from the DOE Passive Solar Program include:

- Building design must be viewed and analyzed as an integrated whole and not simply as an assemblage of individual components and design features.

- A particular design evolves from the set of building objectives, expected occupancy patterns, local climate, and other site-specific factors.

- The most effective designs are simple.

- Daylighting plays a major role in the overall energy system design.

- Good interactive and analytical tools are lacking for the design of commercial buildings.

- Interaction among different members of the design team is crucial (e.g., architects and HVAC engineers need to work together).

In addition to the evaluations of first cost and energy performance, the program included surveys of building managers and occupants (Gordon et al. 1985). Based on month-by-month reports from 15 buildings, occupant satisfaction was very high. Also, occupants were generally pleased with thermal comfort, lighting, and air quality in the buildings. A few complaints were registered about acoustics (excessive noise) and thermal comfort (either cool mornings during the winter or overheating during the summer).

The user survey also showed that actual occupancy patterns differed significantly from those predicted. This variation was thought to have a strong influence on energy use, although no attempt was made to determine the actual impacts. The evaluation also showed that building occupants need to understand how the passive elements of buildings operate and what is expected of them for the buildings to operate up to their potential.

C. COST-EFFECTIVENESS OF ENERGY SAVINGS IN NEW BUILDINGS

The BEPS redesign effort and the DOE passive solar program indicate that it is cost-effective to substantially increase the energy efficiency of new commercial buildings. Other studies confirm this conclusion.

Energy consumption and related data have been collected for a substantial number (but not a statistically representative sample) of new

commercial buildings as part of the BECA data base developed by LBL
(Wall and Flaherty 1984). Most of the buildings are "energy efficient."
However, some energy intensive buildings are also included.

Figure 7.6 plots primary energy consumption against first cost for 38
buildings in the BECA data base. The data points are scattered, and
decreasing energy intensity does not correlate with increasing construction
cost. The LBL analysts conclude that an energy efficient building can be
produced over a considerable range of construction costs (Wall et al.
1984).

Some economic analysis was also carried out as part of the BEPS
redesign effort. In particular, the life-cycle costs of three buildings were
studied as the designs were improved. The analysis showed more-favorable
life-cycle costs for better efficiencies. For example, a large office building
located in Raleigh, North Carolina, had an end-use energy demand of 71
kBtu/ft² in the original design, 30 kBtu/ft² in the initial redesign (a 6%
reduction in life-cycle cost), and 24 kBtu/ft² with further modifications,

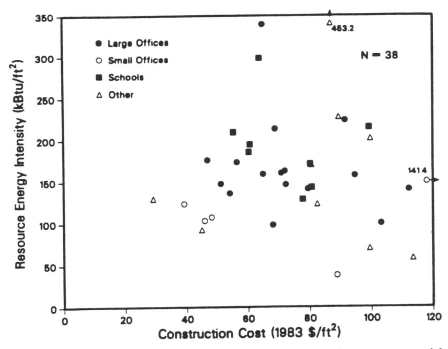

Fig. 7.6. Energy use as a function of construction cost for new commercial
buildings in the BECA data base (Wall et al. 1984).

such as the use of task lighting (a 10% reduction in life-cycle cost) (*Progressive Architecture* 1982b). This analysis is conservative in that low future energy prices and a 10% real discount rate are assumed.

D. SAVINGS POTENTIAL IN EXISTING BUILDINGS

Numerous techniques are available for improving the energy efficiency of existing commercial buildings (Table 7.7). Note the concentration on mechanical and electrical systems. HVAC systems, hot water supply, controls, lights, temperature settings, and control of outdoor air dominate the retrofit categories. Some of these options are discussed further in Section F.

The BECA Group at LBL maintains a data base on energy retrofits in commercial buildings. In mid-1984, they had data on 311 commercial buildings located throughout the U.S., including educational facilities, offices, hospitals, retail facilities, hotels, post offices, restaurants, and assembly buildings (Gardiner et al. 1984).

By far the most frequently cited retrofit strategy (Table 7.8) is improving operation and maintenance of buildings (66%). Modifications to HVAC systems are the next most frequent (44% of the buildings), with lighting modifications third (39%). The extent to which retrofit measures in the BECA data base are representative of retrofit activities in the commercial sector overall is not known.

Buildings in the BECA data base show an average reduction in actual site energy consumption of 27% following retrofit. This value corresponds to an average saving of 45 kBtu/ft^2 (Table 7.9). Assembly buildings realized the greatest average saving (52%), followed by university buildings (41%). In office buildings, the average saving ranged from 24% (large offices) to 28% (small offices).

Examination of broader data on commercial building retrofits shows that the gap between identified energy savings potential and actual savings is often still quite large. For example, Pacific Gas and Electric Co. (PG&E) maintains a data base on its commercial customers that received PG&E energy audits. Approximately 8000 buildings were audited through 1983 (one-third of the commercial building stock in the service area). Based on billing data and a review of the conservation actions implemented, PG&E estimated the amount of savings achieved 18 months after the initial audit (Schultz 1984). Table 7.10 shows the overall results for 5000 buildings. The savings potential estimates are based only on measures recommended in the utility audit (3- to 5-year payback or less) and do not represent the entire savings that might be possible or feasible. The estimated savings achieved are averages for all buildings, whether or not they implemented any of the recommended retrofits. (The extent to which the recommended measures were actually installed is not reported but is expected to be rather low.) The average savings in this very large data set are only 8% for electricity and 6% for natural gas.

Commercial building retrofit information collected by PG&E and other utilities in California shows that lighting efficiency improvements typically

Table 7.7. Retrofit options for commercial buildings

Mechanical system retrofits	Envelope retrofits
Replace burner and controls	Insulate roofs/attics
Replace boiler/furnace	Insulate walls
Install vent damper	Install storm windows
Install stack heat reclaimer	Replace single glazing with
Replace electric resistance heater	double glazing
with heat pumps	Weatherstrip windows and doors
Install boiler turbolator	Insulate windows
Install setback thermostats	Install reflective insulation
Install enthalpy control/economizer	Use shading devices
Replace room air conditioners	Affix roof sprays
Replace central air conditioning	
Vary chilled water temperature	Hot Water Retrofits
Convert terminal reheat to	Install summer domestic hot
variable air volume	water boiler
Reduce ventilation volume	Provide flow control devices
Install evaporative cooling system	Insulate hot water storage
Replace air-cooled condenser	Provide vent damper on heater
with water cooled	Install hot water heat pump
Install fog cooling (evaporator	Reclaim refrigeration heat for
coil spray)	hot water
Insulate ducts	
Insulate pipes	Lighting retrofits
Install two-speed fan motors	Replace incandescent light with
Provide adjustable radiator vents	fluorescent
Reduce orifice size on	Use low wattage task lighting
furnace/boiler	Use high-efficiency fluorescent
Install multifuel boiler	lamps
Use condenser coil spray	Maximize use of daylighting
Install chiller bypass system	Install more efficient lamp
	ballasts
	Install lighting control systems
	General retrofits
	Install energy management systems

Source: Office of Technology Assessment (1982).

account for 50 to 90% of the electricity savings achieved (Kowalczyk 1985). The importance of lighting retrofits may be due in part to incentive programs offered by the utilities during the same period. Boiler and HVAC system modifications provided the bulk of the gas savings.

In Chicago, community-based social service agencies are participating in a retrofit program sponsored by the Neighborhood Nonprofit Energy Program (NNEP). The buildings being retrofitted are predominantly low-rise masonry buildings, which are typical of the nation's commercial building stock. An average energy savings of 31% was obtained with the first ten buildings receiving major retrofits (Katrakis and Becker 1984). The retrofit strategies included adding insulation, using storm windows, installing destratification fans, replacing incandescent with fluorescent lights, improving the control and efficiency of the heating plant and

Table 7.8. **Retrofits implemented in commercial buildings listed in BECA data base**[a]

Category	Percent of buildings
Operations and maintenance	66
HVAC System	44
Cooling	19
Ventilation	28
Control systems	38
Lighting	39
Shell	15
Insulation	9
Caulk/weatherstrip	8
Windows and doors	12
Water heating	10
Other	16
Heat recovery	1
Load management	3

[a]Totals may not add since, in many buildings, more than one measure was installed.
Source: Gardiner et al. (1984).

domestic hot-water systems, replacing boilers and burners, and installing programmable indoor thermostats. In the majority of cases, predicted energy savings were within a few percentage points of actual savings.

Monitored performance data are available for three retrofit buildings in the DOE Passive Solar Program. Those data show that, in gross terms (not correcting for weather or occupancy variations), energy consumption was reduced 46% on the average (Gordon et al. 1985). In addition, surveys found that occupants were generally more pleased with comfort conditions and building operations following the retrofits.

Ohio State University completed a 10-year energy management program involving more than 50 major buildings. The program produced a 40% reduction in average energy consumption and an energy cost

Table 7.9. Summary of energy savings, retrofit cost and cost effectiveness for commercial buildings in the BECA data base

Building type	N^a	Site energy savings		Retrofit cost		Simple payback period	
		Average energy savings (kBtu/ft^2)	Average percentage savings (%)	N	Average cost (1983-$/ft^2)	N	Average payback (yrs)
Schools							
Elementary	82	33.0	27.0	25	1.09	21	5.3
Secondary	28	34.5	27.4	10	2.73	10	1.1
Colleges	13	166.6	41.4	10	1.60	8	0.8
Offices							
Large	34	43.4	24.4	22	0.73	19	1.8
Small	13	52.6	28.5	5	3.26	4	3.4
Hospitals	6	113.8	26.7	5	1.13	5	2.9
Retail	2	8.0	12.3	2	0.02	2	0.2
Hotels	5	51.7	20.3	1	0.47	1	0.5
Other							
Post offices	92	31.2	25.0	93	0.42	59	1.7
Assembly	2	188.6	52.2	2	4.12	1	0.8
Correctional	1	8.5	6.0	0			
Other	14	72.8	27.9	4	0.59	4	0.9
Total	292	45.3	26.7	179	0.75	134	2.2

[a]N is the number of buildings for which the particular type of data is available.
Source: Gardiner et al. (1984).

avoidance during the 10 years of $56 million (Sullivan 1984). These savings were achieved with a capital investment of less than $7 million. The retrofits Ohio State University considers most productive include:

- Limiting operation of equipment to times necessitated by occupancy

- Conversion of reheat and dual-duct HVAC systems to variable volume systems

- Replacement of inefficient steam and hot water absorption chillers with efficient electric-drive centrifugal chillers

- Temperature-control upgrading and use of heat recovery systems

E. COST-EFFECTIVENESS OF ENERGY SAVINGS IN EXISTING BUILDINGS

The Ohio State University example suggests that commercial building retrofits are a very good investment. Other experiences confirm that this is the case. Table 7.9 shows an average retrofit cost of $0.75/ft^2 for buildings in the BECA data base; retrofit capital costs are paid back in 2.2 years on the average (Gardiner et al. 1984).

Retrofit and consumption data from buildings audited by the California utilities showed an average payback of *less than one year* for medium-size and large buildings (Kowalczyk 1985). For very small buildings, average payback periods were less than three years. Surprisingly, the measures implemented most rapidly were not necessarily the most cost-effective (Kowalczyk 1985).

In the Chicago program mentioned above, the average payback period was 5.2 years (Katrakis and Becker 1984). The longer payback in the Chicago program is most likely caused by the greater percentage of envelope retrofit measures than in the retrofits cataloged in the BECA data base.

Although virtually all data on commercial building retrofits show a high level of cost-effectiveness, considerable scatter in energy savings, cost, and cost-effectiveness occurs among buildings. This variation is understandable given the diversity of the commercial sector, the peculiarities of individual buildings, and the sensitivity of energy consumption to building operation.

F. SAVINGS POTENTIAL IN BUILDING EQUIPMENT

Table 7.7 shows that a wide array of technologies is available for reducing energy and power consumption in commercial buildings. This section examines some of the specific products and design concepts developed during the past 10 years.

Table 7.10. **Summary of energy consumption and savings in PG & E's**
large-scale commercial building audit program

End-use	Pre-audit site energy consumption (kBtu/ft^2)	Identified savings potential[a]		Savings accomplished[b]	
		(kBtu/ft^2)	(%)	(kBtu/ft^2)	(%)
Electric					
Space heat	0.4	0.1	25	0.0	0
Air cond.	8.6	1.6	19	0.6	7
Ventilation	2.3	0.3	13	0.1	4
Hot water	0.4	0.2	50	0.0	0
Lighting	17.3	5.1	29	2.1	12
Cooking	1.1	0.1	9	0.0	0
Refrigeration	6.6	0.9	14	0.3	5
Other	3.6	0.1	3	0.1	3
Subtotal	40.3	8.4	21	3.2	8
Natural Gas					
Space heat	26.2	3.3	13	2.1	8
Hot water	6.7	1.4	21	0.4	6
Cooking	5.5	0.3	5	0.2	4
Other	7.6	1.2	16	0.2	3
Subtotal	46.0	6.2	13	2.9	6
Overall total	86.3	14.6	17	6.1	7

[a]Based on PG&E audit.
[b]Eighteen months after initial audit for all buildings audited.
Source: Schultz (1984).

Top-rated commercial building cooling systems now on the market have efficiencies 50 to 100% greater than those in current use (Usibelli et al. 1985). VAV systems are particularly efficient for heating and cooling (Int-Hout 1984; MacDonald et al. 1982). Such systems reduce the amount of air that must be conditioned and the fan power needed for air circulation by lowering overall ventilation rates. Simulation studies show that VAV systems can reduce energy consumption for cooling by up to 50% and energy consumption for heating by an even greater percentage (Parken et al. 1982).

"Economizer" cycles also reduce HVAC system energy consumption. With these systems, unconditioned outdoor air is used for cooling when the temperature or enthalpy of ambient air is low enough. (Cooling interior areas is often required year-round in commercial buildings because of high internal heat gains from occupants and equipment.) Economizer cycles can reduce cooling energy consumption in office buildings by 10 to 50%, depending on the location (Parken et al. 1982).

Heat pumps can lower energy consumption in commercial buildings, particularly when substituted for electric resistance heating. Shipments of commercial heat-pump systems (under 15 tons in capacity) increased from around 20,000/year in the early 1970s to 70,000/year in 1980 (MacDonald et al. 1982).

Energy management systems (EMSs) can be used to control HVAC and lighting equipment as well as to monitor equipment operation in commercial buildings. EMSs range from simple point-of-use timeclocks to full-fledged computer systems. Following the introduction of microprocessor-based EMSs in the mid-1970s, the application of EMSs expanded rapidly. A survey of 200 commercial buildings in 1984 showed that 70% of buildings larger than 50,000 ft^2 are using EMSs. Computerized control systems typically provide a 10 to 20% energy savings (Novey 1984; MacDonald et al. 1982). EMSs can also limit peak electrical demand by selectively switching or cycling certain loads. Of course, the amount of savings depends on the particular application and the ability of building personnel to operate the system (Usibelli et al. 1985).

Controls for HVAC systems are also being improved (MacDonald et al. 1982). Such controls can schedule equipment, vary motor speeds to increase partial-load efficiencies, vary air or water volumes and temperatures, and adjust dampers.

Numerous techniques and improved products are available for reducing lighting energy consumption in commercial buildings, including:

- Replacing incandescent bulbs with other lamp types,

- Using more-efficient fluorescent lamps and ballasts,

- Using high pressure sodium lamps,

- Using task lighting,

- Using daylighting, and

- Using personnel sensors and other controls.

Among the new lighting technologies that have been introduced in recent years are compact fluorescent and other high efficiency lamps that consume up to 75% less power than incandescent bulbs with equivalent light output. For ordinary fluorescent lighting, new ballasts were introduced in the early 1980s that reduce electricity consumption by 20 to 35% relative to standard fluorescent lighting. For indoor applications, such as warehouses or hallways where light quality is not critical, high-pressure sodium lamps provide a 45 to 60% saving compared to mercury-vapor or fluorescent lighting (Usibelli et al. 1985).

A variety of lighting controls (including daylighting sensors and controls, scheduling controls, and occupancy sensors and controls) are commercially available and are gaining in popularity. These controls better match light output to the visual needs of building occupants, reducing electricity consumption by up to 50%. By using more efficient hardware in combination with daylighting and scheduling controls, electricity savings of 60 to 80% can be realized in fluorescent lighting applications (Verderber and Rubinstein 1984). Decreased lighting energy consumption also lowers heat gains, which in turn cuts cooling requirements. This result can be especially valuable for reducing peak cooling loads (Andersson et al. 1984; Selkowitz et al. 1984).

G. BARRIERS TO INCREASED ENERGY EFFICIENCY

Few building owners are making the full range of efficiency investments even though life-cycle cost analyses show such improvements to be economical. Many factors contribute to this situation (Table 7.11).

Most of the barriers leading to suboptimal investments in energy efficiency in commercial buildings are related to individual or institutional behavior (MacDonald et al. 1982). These barriers include reluctance to make perceived-as-risky investments, confusion about the impact of specific efficiency measures, concern for nonenergy aspects of the work environment, and managerial requirements. The expected savings from an efficiency improvement is by no means the only consideration that affects decision-making. The construction and operation of an office building, for example, represent a small percentage of the total cost of the activities a building accommodates. The annual cost to an employer for a typical office worker is approximately $150/ft^2 (Int-Hout 1984), while energy costs are typically $1 to $2/ft^2 in office buildings. Therefore, risking a decrease in worker productivity for the sake of lower energy bills would be a mistake. Even the fear of lowering productivity can prevent designers, owners, and operators from installing efficiency improvements.

Table 7.11. Common barriers to energy efficiency improvements in commercial buildings

Capital

- Investments in increasing efficiency of facility are of low priority.

- Regulatory and political costs of borrowing are too high to justify using debt capital (public sector and non-profit organizations).

- Organization does not have access to internal or external capital sufficient to cover project costs.

Risks

- Building owners are skeptical of performance claims for energy efficiency technology.

- Building owners cannot make independent judgment of performance claims.

- Building owners and occupants are concerned that efficiency improvements will lower worker productivity and have other negative side effects.

Management

- Building owners believe they cannot afford time to manage planning and implementation of a project.

- Staff need to be trained to adequately maintain and operate new energy-efficient technologies.

- Building owners pass through operating costs to tenants.

- Buildings are centrally metered and tenants are prevented from making modifications.

Source: New York State Energy Research and Development Authority (1984).

The nature of building ownership and the manner in which energy costs are paid can inhibit investment in energy efficiency, particularly in such commercial subsectors as retail facilities and large office buildings. In these areas, a high turnover rate in ownership occurs, and the owner tends to pass through the energy bills to the building occupant. In either case, little incentive exists for the owner to invest in efficiency improvements unless the payback is very rapid. The disincentive to invest in efficiency in certain commercial subsectors is confirmed by the BECA data base, where the average retrofit expenditure in retail and large office buildings is much lower than in assembly buildings and schools (Table 7.9).

In situations where tenants are individually metered, lease agreements may restrict what the tenant can do to modify any part of the building. Thus, tenants are prevented from acting in their own interest by cutting energy costs.

Other factors can inhibit energy conservation. First, space conditioning and control systems are becoming more and more complicated. To work, these systems must be installed as designed without any corner cutting on the part of the building owner or HVAC contractor. Second, a building might have to be carefully tested to "debug" sophisticated equipment, such as computerized controls (Warren 1984). Third, operating and maintenance personnel need to be trained to use energy conservation equipment effectively. They must resist the temptation to minimize complaints from tenants by raising general ventilation rates, permanently overriding controls, etc. Lack of attention in these areas can compromise or negate efforts to improve efficiency (Warren 1984).

Experience with new buildings as well as retrofits has shown that the accurate prediction of the energy impact of any particular efficiency improvement is difficult. Although a building's energy performance can be modeled through computer simulation, too many variables exist to have a great deal of confidence in the validity of the simulations. The unpredictability of climate, behavior of occupants, changes in tenants during the analysis period, and the building's operation and maintenance not only produce large discrepancies between predicted and actual energy use but also discourage efficiency investments.

H. UNRESOLVED ISSUES

Some of the unresolved issues related to energy efficiency in commercial buildings have been touched upon already. Additional information and research is most needed in the following areas (Harris 1984):

- More research and data are needed concerning the general progress of energy efficiency in commercial buildings. This need implies further building monitoring, including following the same buildings for several years. Publication of the *1983 EIA Nonresidential Buildings Energy Consumption Survey* should help meet this need. The EIA survey could be more useful if it included more detailed information on retrofit activities. Also, data from suppliers on equipment sales and retrofit activity would be useful.

- Conservation activities, particularly in the retrofit area, seem to be strongly influenced by a "one-time technical fix" mentality. Emphasis on continuous monitoring of the performance of buildings and periodic retrofits should help identify operational problems, attract the attention of owners/managers, and counter the one-time technical fix mindset. Along similar lines, training or economic incentives may be useful to those operating larger commercial buildings.

- At the individual building scale, most of the monitoring efforts to date measure whole-building performance. Very little data exist on end-uses (heating, cooling, lighting, etc.) or on the separate contributions to energy savings of individual measures where a multiple set of retrofits have been implemented. Also, changes in electricity load profiles have rarely been well documented, even though reduction of peak demand and shifting of electrical use to periods with lower rates are of major importance to utilities and commercial building owners.

- Research on the broader implications of increasing energy efficiency in commercial buildings is still at an early stage of development. More work is needed on how highly efficient new buildings and major retrofits affect comfort, indoor air quality, worker productivity, and other qualitative issues, such as aesthetic appeal. The possible increasing concentration of indoor air pollutants is of particular concern with HVAC strategies that reduce flow rates. (See Chapter 11.)   On the other hand, some measures that improve energy efficiency, such as daylighting or atria, can provide a better working environment both visually and aesthetically (Harris and Blumstein 1984).

- Reducing lighting energy consumption in commercial buildings leads to additional issues, such as the optimal lighting levels for different tasks and the provision of lighting systems that are efficient and flexible. Research is also needed on defining lighting quality and comfort, considering factors such as intensity, distribution, glare, and color.

- How nonequipment factors, such as climate and the behavior of the user or operator, affect whole-building energy consumption is another problem. This issue complicates year-to-year interpretation of energy consumption data and is also a practical concern in designing and using more efficient buildings.

- How to affect the behavior of different actors associated with energy consumption in commercial buildings is another major problem.

Techniques for stimulating greater concern for and investment in efficiency need to be developed, studied, and disseminated. Accurate yet simple design tools for architects are lacking (Gordon et al. 1985). Strategies for overcoming resistance to efficiency investments on the part of building owners need to be systematically analyzed.

- Several important concerns exist regarding advanced technologies. First, controlled experiments should be conducted by independent organizations to document savings. Second, more information on new equipment and design concepts should be disseminated to accelerate market penetration. Finally, R&D on promising technologies not yet commercially available should continue, including work on advanced thermal storage techniques, new window coatings, advanced lighting products, and improved refrigeration mixtures.

8

INDIVIDUAL AND INSTITUTIONAL BEHAVIOR

A. INTRODUCTION

An oft-repeated truism among energy conservation analysts is "Buildings don't use energy, people do!" Although much of the discussion, here and elsewhere, concerns the technical performance of various energy conservation devices and designs, it is essential to remember that people decide what, when, where, and how much energy is consumed. Obviously, building occupants influence energy use through various behaviors, such as setting temperatures, opening and closing windows and doors, and using hot water. People also affect energy use in other ways—as individuals and as members of organizations that commission, design, construct, and manage buildings and equipment.

The first part of this chapter discusses energy efficiency *purchase* decisions (e.g., the process of deciding between an energy efficient or an "electricity-guzzling" refrigerator). The second part discusses individual behavior related to *operation* of buildings. More is known about the behavior of individuals and households in residences than about building managers and occupants of commercial buildings. The final part deals with the behavior of organizations (such as builders and architects, equipment manufacturers and suppliers, financial institutions, and associations) involved with building design and construction.

B. CONSUMER PURCHASE BEHAVIOR

Choices between models or methods in the purchase of appliances, heating and air conditioning equipment, new homes, and retrofit measures for existing homes can influence future energy consumption.

The economically rational person, in choosing among various alternatives that meet the same end use (e.g., keeping food cold), will trade off the higher capital cost for a more efficient unit against the lower operating cost over the lifetime of that unit. That tradeoff can be quantified with the concept of the implicit discount rate. The higher the discount rate, the more the individual emphasizes initial costs relative to future savings. In other words, higher discount rates imply less investment in more expensive energy efficient systems. The term "implicit" refers to the fact that most analyses infer discount rates from household behavior; they do not rely on individual self-reports of discount rates.

Several econometric analyses have been conducted during the past several years to statistically examine this relationship. The seminal work by Hausman (1979) analyzed purchase of room air conditioners. It showed a very high implicit discount rate for these purchases, about 24%. Interest rates varied inversely with household income: households with annual incomes below $10,000 had discount rates of 40% or more, whereas households with incomes above $35,000 had discount rates below 9%. Other economists (Dubin 1982; Goett 1983) have applied models similar to Hausman's to estimate implicit discount rates for purchase of energy efficient residential equipment.

Analysis of aggregate data on appliance purchases from 1972 to 1980, based on national appliance shipment data, shows discount rates that are very high (ranging from 10 to 200%) and roughly constant over this period (Ruderman et al 1984). Such analyses, based either implicitly or explicitly on economically rational models of human behavior, generally show discount rates much higher than real interest rates or those commonly used in public decisionmaking (Fig. 8.1). These findings indicate that households generally underinvest in new energy efficient space heating equipment and appliances. Several government and utility conservation programs encourage increased production and purchase of high efficiency equipment; see Chapters 9 and 10.

Unfortunately, these analyses provide few insights into particular barriers that inhibit purchases of efficient equipment. Implicit discount rates reflect many factors: the types of products offered by manufacturers, the types and usefulness of information available to purchasers, the time required to obtain and process information, the availability of capital, and who makes the purchase (e.g., builder or homeowner). Stern (1984) concluded that to "subsume all these influences under a single index and call it the discount rate may be to misconceive the phenomena . . . such theoretical shorthand may lead analysts to think of some features of energy users' behavior as stable when they may in fact be changed by economic or institutional forces or by policy."

In actuality, people do not think in terms of the time value of money (Feldman 1984). When surveyed, most people knew, at least

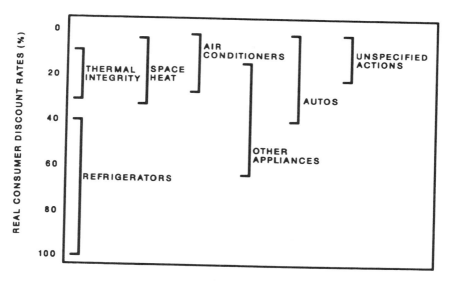

Fig. 8.1. Estimates of average discount rates used by households in making energy-related purchase decisions (Train 1985).

approximately, last year's inflation rate, but they were unable to determine "how money functions as a standard of deferred payment." When asked "how much they would have to spend today to get what one dollar bought a year ago [1983]," the median response "was an astounding $1.41," implying an inflation rate perception an order of magnitude greater than the actual inflation rate.

Surveys of households in Santa Cruz, California, (a decidedly atypical sample of U.S. households) showed that the conservation actions taken vary "markedly between homeowners and renters, and across different types of conservation" (Archer et al. 1984). Installation of conservation devices was positively related to socioeconomic status and to the availability of a household member able to do home repairs.

A similar survey conducted by Wilk and Wilhite (1984) in Santa Cruz examined why people do not caulk and weatherstrip doors and windows. Because people viewed these measures more in the realm of home repair than in the realm of home improvement, these initiatives lacked the glamour of other more visible measures, such as storm windows and solar water heaters.

Households have inaccurate perceptions of the major energy users in their homes (Kempton et al. 1984). As a consequence, people overestimate the energy savings produced by management and curtailment practices (e.g., turning off lights) and underestimate the potential of energy efficiency investments. These incorrect perceptions of energy use and conservation options help explain why households "underinvest" in efficiency measures and why many government and utility conservation programs focus on encouraging such investments.

C. CONSUMER OPERATING BEHAVIOR

HUMAN FACTORS

Many studies show that occupants have substantial effects on building energy use. For example, only about half the house-to-house variation in winter gas consumption for townhouses in Twin Rivers, New Jersey, could be explained by physical characteristics, such as the number of bedrooms and area of insulated glass. Almost three-fourths of the remaining variation was caused by "occupant-related consumption patterns" (Sonderegger 1978).

Indoor temperatures can have a considerable influence on residential energy consumption. A reduction in indoor temperature from 70° to 68°F will cut annual space heating energy use by 10% in locations with 5000 heating degree days (65°F base). Reducing the nighttime temperature from 68° to 60°F (for eight hours each night) increases the overall energy saving to 23% (Pilati 1975). Many analysts believe that much of the post-embargo decline in per-household energy use was caused primarily by behavioral changes.

Although occupant behavior substantially affects energy use (Fig. 8.2), our understanding of the factors that influence such behavior is limited. In fact, conventional wisdom concerning the motivation for these behaviors is probably incorrect. According to Stern and Aronson (1984), "Most analyses proceed from the simplifying assumption that energy producers and consumers are rational economic actors: that is, that they are motivated to maximize the value of some objective function, such as income, profit, or organizational size." Although in aggregate this assumption may be useful, it generally obscures the large variation across energy users.

Households might behave in apparently irrational ways because the majority of a typical household's energy use is automatically controlled and *invisible* to its members. Homeowners are therefore unaware of how much energy they use for which functions. (See discussion in Chapter 5 of Kempton's findings on inaccurate perceptions of household energy users.)

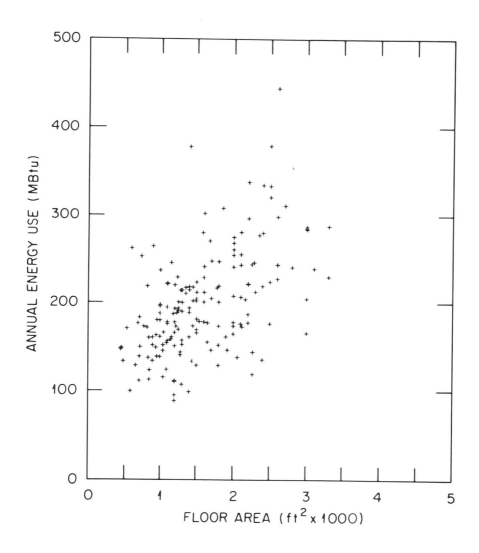

Fig. 8.2. Energy use as a function of floor area for single-family homes heated with natural gas, built after 1950, in cold northern climates (6000 to 8000 HDD). Note the large variation across homes of similar size, probably due primarily to differences in occupant behavior.

In addition, households might adopt seemingly irrational behavior because they mistrust generalized information sources. Because people may realize that energy saving prescriptions appropriate for one house are not necessarily suitable in another, they may refuse to install some recommended measures.

The different ways in which people behave yield substantial diversity in energy use. Stern and Aronson (1984) offer five models of the individual as energy user. The first is the conventional economically rational user—*the investor*—who makes appropriate tradeoffs between the operating and capital costs of various energy efficiency choices.

The second is the *consumer*, who thinks of energy using possessions as providers of necessities and pleasures, focusing primarily on the benefits obtained with little regard to the economics of ownership and use.

The third is the member of a *social group*. In this model, households are influenced primarily by friends and neighbors both in purchase decisions (e.g., the kind of car to buy) and in behaviors (e.g., at what temperature to set their thermostat).

The fourth relates energy use to *personal values*; energy consumption is a consequence of one's values and self-image. For example, those who are concerned about environmental quality may be frugal energy users, regardless of the direct economic benefit of saving energy. Those who are proud of a comfortable, affluent lifestyle may keep their houses cold in summer and hot in winter.

The final model is the *problem avoider* for whom attention is a scarce resource. Energy efficient investments are not made and energy conservation practices are not adopted until some threshold is reached (e.g., a particularly high heating bill).

Stern and Aronson concluded that none of these models is correct. Rather they all contain some elements of truth, the amount varying from household to household. These alternative models suggest that the economically rational model may not always be correct and—more important—may lead to policies and programs that are ineffective. For example, if one considers households as investors (the first model) whose supply of capital resources (money) is limited, then a tax credit for retrofit investments seems attractive. On the other hand, if one considers households as problem avoiders, then programs that offer convenient ways to improve the energy efficiency of their homes may be more effective than a tax credit.

SYSTEM FACTORS

The physical systems in individual homes and the options that households actually face in managing these systems should be examined as well as household attitudes and demographic factors. Management of energy using systems depends at least partly on the systems themselves. Thermostat management (especially nighttime thermostat setback) is much easier to implement in a home with a central heating system and one thermostat than in a home with individual room heaters. Load research data from a sample of homes in Minnesota shows lower electrical loads at night and higher loads in the morning for one-thermostat homes relative to a group of homes with individual room heaters (Fig. 8.3; Kuliasha et al. 1985).

Fagerson's (1984) analysis of new energy efficient homes and retrofit homes, also in Minnesota, has similar findings. In addition, thermostat settings were closely correlated with the amount of south-facing window area and with infiltration rates. Thus, people in leaky homes raise temperatures more than those in tighter homes do.

In multifamily buildings, the limitations of existing systems play a vital role in energy efficiency. Reports abound of households that open windows in winter if the central space heating system provides too much heat. They do so because it is the only way they can control temperatures within their

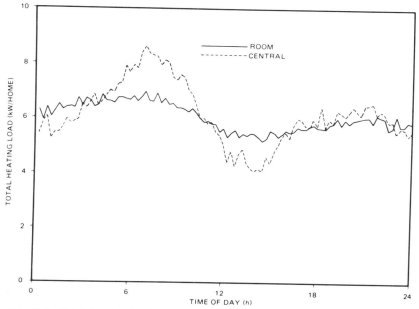

Fig. 8.3. Total heating load for electrically heated homes with central heating vs homes with individual room heaters (Kuliasha et al. 1985). The data strongly suggest that families with central heating systems reduce temperatures at night.

apartment. On the other hand, some people use gas stoves to provide space heat when the central system does not provide sufficient heat.

SOCIAL SCIENCE FACTORS

One of the difficulties in assessing the importance of household behavior is the traditional reliance on self-reports. In many cases individuals either cannot or will not provide correct answers to questions concerning their behavior. A recent evaluation of a utility home energy audit plus zero-interest loan program found that 4% of the households that received both an audit and a loan claimed to have received neither; 16% of the households that received an audit only claimed not to have been audited; and households that took loans reported only 79% of the utility-certified retrofit measures that had been installed (Hirst and Goeltz 1985).

In part because of errors in household self-reports (as well as a desire to better understand the actual patterns and behavioral determinants of household energy use), a group at Michigan State University (see Kempton 1984; Kempton and Krabacher 1984; and Weihl 1984) is collecting engineering data (thermostat settings, actual indoor temperatures, hot water consumption, etc.) and behavioral information (from household questionnaires and open-ended ethnographic interviews). Although the number of homes being monitored in this project (less than ten) is far too small to draw statistically valid conclusions, the research already points to some important findings. Energy use patterns (1) are different on weekends than on weekdays (particularly important with respect to electric utility peak loads), (2) depend on the coincidence of schedules for different family members (e.g., whether they eat meals together), and (3) depend more on the number of adults than on the number of children in the home. Households that have few family members, that have regular bedtimes, and that are usually not home during the day have the most regular energy use schedules. The MSU researchers hope that careful examination of their detailed data will suggest which are the few key variables needed to understand household energy related behavior and to design effective conservation programs.

Analyses of summer air-conditioning use in two California communities focused on development of econometric models of household energy use (Cramer et al. 1984; Vine et al. 1982). They developed a two-part model of summer electricity use. The first part estimated electricity use as a function of the physical factors that directly determine energy use: air conditioners and electric appliances. The other part estimated appliance ownership, house size, number of rooms closed, and frequency of air conditioner use as functions of economic and demographic factors and of household attitudes towards comfort, energy conservation, and environmental issues. Family income and the number of household members were the most important indirect determinants of energy use. The California approach is much less expensive and much easier to implement than the MSU approach, but is less likely to yield detailed understanding of household behavior and its interaction with residential energy using systems.

COMMUNICATION FACTORS

A careful review of 200 evaluations of energy conservation programs conducted by California electric and gas utilities suggests that traditional generalized advertising approaches (e.g., leaflets and radio and TV ads) are likely to be ineffective ways to get people to change their patterns and to promote energy conservation (Condelli et al. 1984). Instead, existing social networks (e.g., community leaders and local nonprofit groups) should be used to spread information about effective conservation actions, and information should be made vivid and personalized. For example, having the energy auditor caulk one window is probably a much more effective way to encourage adoption of other measures than is a detailed written description of the likely economic benefits of different measures. Similarly, reference to the experience of other households in the same community is more effective than general statements about benefits.

One way to make energy savings visible and vivid is to provide feedback—regularly telling residents how much energy they use. A number of experiments, summarized by Sorenson (1985), show that the amount of energy saving achieved depends on the frequency of feedback. The studies also show that the combination of feedback and provision of energy conservation information is even more effective than either alone (Table 8.1). Provision of daily feedback is probably not currently feasible without installation of new devices that monitor and present energy consumption information. However, current and emerging microprocessor technologies may soon permit installation of low-cost systems in homes. These systems may perform energy management functions (as do currently available, computerized, energy management systems for large commercial buildings). In addition to providing more cost efficient heating and cooling, these systems can provide feedback to households on their energy use (and cost). For the near term, gas and electric utility bills could be modified to provide feedback on changes in energy use (from month-to-month, from this month to the same month last year, or from year-to-year) with little difficulty (Fels and Kempton 1984).

Table 8.1. Estimated energy savings from information

Type of information	Savings[a] (percentage)	
	Average	Range
Energy savings goals	1	0–1
Information	4	0–9
Feedback on consumption	11	3–21
Financial incentives	15	4–28
Combined feedback and information	14	13–17

[a]The large range in estimated energy savings for similar programs illustrates the variation in the implementation and consequent effects of those programs.

Source: Sorensen (1985).

Improving operating practices in multifamily buildings is much more difficult than in single-family homes because it involves both owners (or their managers) and tenants. In addition, the central heating equipment and associated distribution system are more complicated in multifamily buildings, especially high-rise buildings. Wisconsin Gas Company's Voluntary Rental Unit Conservation Program includes two sets of energy audits, structural audits intended for building owners and personal energy audits intended for tenants (Fay 1984).

D. INSTITUTIONAL BEHAVIOR

People affect building energy efficiency as members of organizations involved in the design, construction, financing, purchase, operation, and occupancy of buildings. This section summarizes the influences that various organizations and professionals have on energy use in buildings. The roles of designers, builders, appliance manufacturers, and financial institutions are discussed here. The roles of government agencies and energy utilities are discussed in the following two chapters.

DESIGNERS AND BUILDERS

Many organizations make significant contributions to improved energy efficiency in buildings. Some are interested in whole-building performance while others focus on particular products or processes. These institutions represent various segments of the building industry: architects, engineers, building owners and managers, equipment manufacturers, and building-materials manufacturers. These organizations include:

American Consulting Engineers Council
American Institute of Architects
American Planning Association
Air Conditioning and Refrigeration Institute
American Society of Heating, Refrigerating, and Air-
    Conditioning Engineers
Building Owners and Managers Association
Building Thermal Envelope Coordinating Committee
Illuminating Engineering Society of North America
Masonry Industries Committee
Mineral Insulation Manufacturers Association
National Association of Home Builders
National Institute of Building Sciences
Passive Solar Industrial Council
Solar Energy Industries Association
Urban Land Institute

The recent energy efficiency related activities of the American Institute of Architects (AIA) and the National Association of Home Builders (NAHB) are discussed here as examples.

The NAHB has 127,000 members that employ 3 million workers. When the economy is healthy, the average homebuilder produces 20 to 25 homes per year. In an economic slump, production may decrease to 10 homes per year. The membership is organized into 780 local associations, each of which plans, organizes, and manages its own programs, such as energy conservation.

One of NAHB's major energy related activities was its 1979 development of voluntary energy guidelines. The "Thermal Performance Guidelines" were developed for 211 cities in all U.S. climatic regions. These guidelines were made available to local associations for voluntary use in their individual programs and projects.

Pilot programs were organized around these guidelines; they involved seminars, workshops, and conferences. These guidelines have also been accepted by many lending institutions. NAHB feels there is little incentive to go further than the application of the guidelines, which focus on standard energy conservation technologies, because interest rates appear to dominate home purchase decisions.

Both builders and home buyers alike are extremely sensitive to first cost and housing "affordability," producing some corner-cutting on quality including compromises in thermal integrity. Even though "energy efficiency" was viewed as very important by three-fourths of surveyed home buyers in 1983, efficiency depends on perceptions as well as on the structural characteristics of the house itself. Builders and buyers are largely concerned with readily observable features that provide status, sales appeal, and other amenities at the least cost. This clearly works against care in construction and the provision of hidden materials (e.g., wall insulation and vapor barriers) to save energy.

The AIA has 44,000 members working in approximately 12,000 firms. It sponsors numerous energy conscious design programs and its *AIA Journal* often discusses energy-conscious projects. Its AIA Foundation helped develop the Building Energy Performance Standards (BEPS) for DOE.

A recent AIA effort has been its Energy in Design Workshop program for which it developed a multivolume set of design and redesign guidelines. These volumes present energy conscious guidelines from overview, schematic, and detailed perspectives. These volumes have been used in workshops held in various cities thoughout the U.S., open to all building-design professionals.

The AIA Foundation is involved in a series of workshops, sponsored by various government and private organizations. The goal is the formulation of a national agenda for research in energy efficiency and in building-related solar energy.

The AIA has no way of ensuring that its members design efficient buildings. However, evidence from its design awards program, publication of outstanding projects, and attraction of research funds through the AIA Foundation suggests that AIA is a significant force in stimulating progress in energy efficient architecture.

APPLIANCE MANUFACTURERS

The home appliance industry sells approximately 30 million major appliances and space-conditioning systems annually. Even though the appliance industry is concentrated among large domestic corporations (such as General Electric, Whirlpool, and White Consolidated Industries), it is very price competitive. Throughout the 1970s, appliance prices increased at less than half the inflation rate (Science Applications, Inc. 1982). Appliance price rises were moderated by both competition and increasing productivity. Furthermore, appliance manufacturers typically spend only 1 to 2% of their sales revenues on R&D, compared to 4 to 7% of revenues in such industries as electronics and instrumentation (Sterling Hobe 1984).

Little information is available on how manufacturers treat energy efficiency in decisions regarding product offerings. Manufacturers regard such information as proprietory. From the products offered in the marketplace, it can be seen that manufacturers have made major advances in developing and introducing more-energy-efficient products during the past decade (Chapter 6). Assessments of available highly efficient models show that they are still far from life cycle cost minimums. The emphasis in the U.S. appliance industry has been on minimizing price rather than on technological advancement (Sterling Hobe 1984).

Understanding how manufacturers view efficiency improvements (e.g., what paybacks are required to motivate production of more efficient products) would be useful. Another important manufacturing issue is the potential of Japanese and other foreign manufacturers to export efficient appliances to the U.S.

FINANCIAL INSTITUTIONS

Success in achieving energy efficiency often depends on the ability of building owners and users to put up the necessary funds or to link up with an outside party that provides financing. Unlike investments in utility power plants or major oil and natural gas projects, capital commitments to energy efficiency are made on a very small scale and by a myriad of individuals and organizations.

Sources of capital for energy efficiency investments in buildings include internal funds, banks and other conventional lending sources, utility loan programs, government financing programs, and third-party financing. The remainder of this section discusses the behavior of conventional financing sources (e.g., banks) and the expanding area of third parties who provide financing for conservation investments.

Banks and Other Conventional Lending Sources

Traditional guidelines for mortgage lending and housing appraisals discourage builders and homeowners from incorporating energy efficiency components into new housing (Schuck and Millhone 1982). Instead, these guidelines tend to limit home mortgages according to the income of the purchaser without considering operating costs (i.e., energy bills).

In recent years, interest has been shown in providing increased mortgages for highly efficient houses or in adding the cost of a retrofit to the mortgage for a less efficient house. The rationale for such efforts is that it easier for the owner to make a larger mortgage payment if the home is energy efficient. This scheme has been successfully tried in Massachusetts and in Seattle (Kline 1982). Also, the Federal Home Loan Mortgage Corporation (Freddie Mac) and the Federal National Mortgage Association (Fannie Mae) have embraced the concept (Schuck and

Millhone 1984). However, the concept has not been widely implemented by local banks. A 1984 survey of 150 lending institutions showed that 88% do not give preferential loan treatment to people buying energy efficient homes (Owens Corning 1984). Although most of these lenders sell loans in the secondary market, nearly 90% are not aware of the special underwriting guidelines for energy efficiency offered by Fannie Mae and Freddie Mac.

Because such mortgages can provide a major incentive for energy efficiency improvements in housing, additional efforts to change the policies of local banks are called for. Nearly half the lenders surveyed claimed that, in principle, they would be willing to change their debt-to-income ratio for energy efficient homes (Owens Corning 1984).

Bank loans are also used to finance residential energy retrofits and major equipment purchases (e.g., HVAC systems). However, interest rates for such loans are high, so consumers are reluctant to use them. Other borrowers may not be able to qualify under the restrictions placed by banks on these small loans.

Third-Party Financing

Third-party financing, which involves capital investment by someone other than the building owner or the supplier, is rapidly expanding. As of 1984, more than 100 energy service companies (ESCOs) were willing to install and provide financing for energy conservation measures in buildings (Klepper 1984). By and large, ESCOs rely on outside investors to provide funds for their projects (New York State Energy Research and Development Authority 1984). In many cases, ESCOs lease equipment to, operate equipment for, or enter into a "shared savings" plan with the building owner.

The possibility of third-party ESCOs making efficiency improvements in commercial or multifamily residential buildings overcomes a number of the traditional barriers to conservation in larger buildings. These barriers include a low level of awareness and skepticism regarding conservation options, lack of motivation, lack of capital, high required rate of return, and low priority for conservation investments (Hobbs et al. 1984).

Still, several issues and challenges confront ESCOs (Klepper 1984). ESCOs will have to build credibility through demonstrated "success stories" and through information dissemination. So far, third-party financing has mainly occurred in large buildings where conservation measures cost at least $50,000 (Klepper 1984). Further work is needed to reach smaller buildings in a practical and cost-effective manner. In addition, the problem of "cream skimming" (i.e., making only very-short-payback investments, the investments with the least risk) is a concern.

At the institutional level, standard arrangements for energy service agreements and for measuring energy savings need to be developed. Also, the status of tax benefits for third-party financing for conservation investments is uncertain and needs to be stabilized (Klepper 1984; Weisenmiller 1984).

On the positive side, third-party financed efficiency investments are beginning to penetrate hard-to-reach markets, such as public housing, schools, hospitals, and government buildings. Experiences have shown that while negotiations and implementation can be difficult, state and local governments are well-suited for using third-party financing arrangements (Breed and Michaelson 1984; Shinn and Rametta 1984; Weedall 1984). The procedures for energy-service contracting by public sector organizations are expected to improve with time as this market expands.

More recently, the shared savings/energy service idea has been extended to single-family homes on a demonstration basis. Sentinel Energy, a private company, provides energy audits and retrofits in Hennepin County, Minnesota, at no cost to homeowners (*Energy Conservation Bulletin* 1985). Sentinel then keeps 60% of the actual energy savings for five years to repay its capital investment in these houses. Whether or not this strategy works and leads to substantial energy savings is yet to be determined.

9

GOVERNMENT CONSERVATION PROGRAMS

In this chapter we look at the energy program activities that have been developed or carried out by federal, state, and local governments as well as community groups. In many cases these programs involve collaboration between public and private organizations. A common attribute of these programs is that they have been developed to overcome the barriers inhibiting wider implementation of cost-effective efficiency improvements. The programs are assessed in five broad categories: planning activities, information programs, economic incentives, regulations, and comprehensive "energy-service" programs. The R&D programs sponsored by the federal government and some states are noted in earlier chapters.

A. PLANNING ACTIVITIES

FEDERAL PLANNING

The federal government adopted several conservation programs in the mid- to late-1970s (Table 9.1). Taken together, these programs were developed to achieve broad objectives, such as the reduction of oil imports and the weatherization of 90% of the homes in the U.S. by 1985. Individual programs were designed to achieve specific objectives, such as increasing the efficiency of residential appliances, providing energy audits for households, weatherizing the homes of low-income families, or increasing the efficiency and cost effectiveness of federally owned buildings and facilities. Some national planning and analysis is called for in the implementation of these specific programs.

Within the U.S. Department of Energy, a number of planning activities affect federal conservation programs as well as federally funded conservation R&D efforts. The Department's *Annual Report to Congress*

Table 9.1. Major federal legislation related to energy conservation in buildings

| PL 94-163 | *Energy Policy and Conservation Act* | 1975 |

Established appliance labeling and standards programs and the state energy conservation program (requiring conservation planning and implementation at the state level). Also authorized the development of energy conservation programs and standards for federal buildings.

| PL 94-385 | *Energy Conservation and Production Act* | 1976 |

Established the energy conservation standards for new buildings (BEPS) program and the weatherization assistance for low-income persons program. Also amended the appliance standards and state energy conservation programs and authorized the provision of demonstration grants and loan guarantees to stimulate conservation measures in existing buildings.

| PL 95-39 | *National Energy Extension Service Act* | 1977 |

Authorized states to establish energy extension services for technical assistance, educational activities, and demonstration programs.

| PL 95-618 | *National Energy Tax Act* | 1978 |

Established the federal residential conservation and solar energy tax credits.

| PL 95-619 | *National Energy Conservation Policy Act* | 1978 |

Created the Residential Conservation Service (RCS) program and the grant program for schools, hospitals, and local-government buildings. Required the national mortgage associations to purchase loans for conservation improvements and authorized grants and standards for improving the energy efficiency of public housing. Amended the Weatherization Assistance Program and the legislation dealing with the minimum property standards for housing insured by federal housing agencies.

| PL 96-294 | *Energy Security Act* | 1980 |

Initiated the Commercial and Apartment Conservation Service (CACS), the Solar Energy and Energy Conservation Bank, and amended the RCS program.

and biennial *National Energy Policy Plan* provide broad policy direction for federal programs. Within the R&D program, both annual and five-year program plans are developed. The most recent five-year plan identifies opportunities for technological advancement, estimates resource and time requirements in areas where federal support is deemed appropriate, and specifies program priorities and elements (Department of Energy 1984c).

STATE PLANNING

Prior to the advent of the State Energy Conservation Program (SECP), little state-level energy conservation planning was performed. State public utility or public service commissions (PUCs and PSCs) primarily regulated electric and gas utilities; they performed little analysis or planning on ways to help consumers reduce their energy bills. During the 1970s, state energy offices were created to address broader energy needs such as problems related to petroleum fuels, employment, environmental quality, and economic growth.

SECP required states to develop state conservation plans and programs in return for federal funding. The required programs included thermal standards for new buildings, lighting efficiency standards for public buildings, consideration of efficiency in state procurement practices, right-turn-on-red traffic rules, and carpool or public transportation programs. The states were free to propose other conservation programs on their own to obtain supplemental funding provided by the federal government.

The SECP program had a goal of reducing overall energy consumption in each state by 5% by 1980. According to reports prepared by the states, this goal was not achieved, and outside reviewers felt that the savings reported were overstated (General Accounting Office 1982c). In addition, the program was not structured in a way that permitted states to optimally spend the resources provided (General Accounting Office 1982c).

With a few notable exceptions, optional state programs developed under SECP have been sharply curtailed by decreases in federal funding during the 1980s (Sawyer 1985). States continuing programs through nonfederal funding often rely on energy surcharges or taxes on energy-extraction activities (e.g., oil exploration and production). In some states where energy agency activities have declined, PUCs are now taking a more active role in planning and ordering utility energy conservation programs. Florida, Maine, New Jersey, California, Nevada, and Wisconsin are examples of such states.

State (or even regional) planning activities, such as those conducted by the California Energy Commission and the Northwest Power Planning Council, foster development of state information or demonstration programs, regulation of utility conservation activities, and adoption of state energy regulatory programs (e.g., appliance standards).

A few states, such as North Carolina, Florida, Minnesota, and New York, have moved beyond the SECP mandate and are now implementing comprehensive conservation programs. Minnesota state officials develop annual conservation plans and support pilot programs, focusing on areas where private sector action is inadequate (Hirst and Armstrong 1980). Besides developing a state energy plan, New York established a state energy R&D agency that spends about 35% of its budget on conservation projects.

North Carolina established an Alternative Energy Corporation funded by contributions from electric utilities as well as by contracts and grants from others. The corporation works with utilities, industry, universities, and government to increase North Carolina electric ratepayers' knowledge of and activity in alternative energy (including efficiency) (Veigel 1984; Veigel and Lakoff 1985).

Florida passed an Energy Efficiency and Conservation Act in 1980. This law requires utilities to evaluate alternatives to building new power plants, specifies a goal of a 25% reduction in oil use for power generation by 1990, and directs the Florida PSC to adopt energy conservation goals for electricity and natural gas. In 1981, each utility developed a comprehensive plan for meeting the goals set by the PSC. Florida Power and Light, for example, has implemented more than 30 programs for reducing power generation by 8 billion kWh/yr by 1990 (the utility generated 47 billion kWh in 1980). Those savings not withstanding, an even more aggressive "least-cost" state conservation plan has been proposed and evaluated by independent analysts (Capehart et al. 1982); least-cost energy strategies are discussed in more detail in Chapter 10.

The major uncertainties associated with planning efforts like those in Florida concern implementation: Does sufficient political will exist to place supply and demand options on an equal footing? Will concerns for revenue generation in the short-term outweigh the desire to limit demand for longer-term benefits? Can the diverse actors (state policy makers, utilities, other private-sector organizations, and consumers) work together to convert plans into reality. Can planned-for conservation be brought on line with a reasonable degree of certainty?

LOCAL PLANNING

Federal and state governments often encourage local authorities to adopt and execute policies to achieve energy efficient community development. Between 1978 and 1981, the DOE Comprehensive Community Energy Management Program funded the development of local energy-management plans in 16 communities. The planning process documented energy use, identified local issues and major energy users, identified strategies for a desired energy future, and outlined the programs and activities required for implementation (Collins et al. 1985).

The California Energy Commission (CEC) funded several communities in 1976 through 1978 to develop "Energy Elements" as part of each community's general plan. The CEC subsequently published a *Local Government Energy Guidebook* with advice on strategies to consider, sources of data to use in assessing local energy use, references for further analysis, and model plans and policy statements (Baumgardner and Schultz 1981). It also published a report containing model local ordinances for promoting conservation and renewable energy use (Western Sun 1979).

State support for local energy planning in California has had some impact. A 1983 survey of California cities and counties indicated that:

10% had adopted an energy element in their local plans,
17% had adopted energy policies in other forms,
33% had modified ordinances to encourage or require energy conservation in land use or building codes, and
34% routinely review new subdivision developments for solar access (League of California Cities 1984).

A few communities are beginning to move from piecemeal programs to comprehensive conservation plans. Austin, Texas, for example, initiated a wide-ranging program to reduce municipal power demand by 550 MW (about 20%) by 1997 (City of Austin 1984). The Austin plan was developed jointly by government authorities and the local municipal electric utility. This plan represents a significant step forward in integrated local conservation planning, but pertains only to electricity.

B. INFORMATION PROGRAMS

Information programs have been the most prevalent method of encouraging improved energy efficiency. The underlying premise is that if energy users know:

• How they are using energy,

• What options exist to improve the efficiency of their use, and

• What benefits they might receive from exercising these options,

then they are likely to take "appropriate" action.

Literally thousands of energy information programs were initiated during the past ten years. The major types of information programs are described here.

FEDERAL PROGRAMS

In the buildings sector, the federal government has three major information programs. These are: Appliance Efficiency Labels, the Residential Conservation Service, and the Energy Extension Service. Because it is implemented at the state level, the Energy Extension Service is discussed with state programs.

Appliance Labels

The Energy Policy and Conservation Act required the National Bureau of Standards (NBS) to develop standard test procedures and the Federal Trade Commission (FTC) to develop and implement energy performance labels for appliances and HVAC equipment. The FTC ruled that energy labels (Fig. 9.1) must be placed on seven types of products: refrigerator-freezers, freezers, room air conditioners, clothes washers, dishwashers, and electric and gas water heaters. Besides providing the estimated annual energy cost for the particular model, the labels indicate the highest and lowest energy costs for models with similar features. For room air conditioners, the labels provide the efficiency rating and the estimated annual operating cost as a function of electricity price and usage.

The current label formats have several shortcomings: some of the major product types are not covered; the labels do not identify the most efficient products available or the relative cost effectiveness of particular models; the labels use national prices that may substantially misstate local costs; and the labels may be hard for some consumers to understand (Geller 1983). Future evaluations should address the question of how label information can be made more useful to consumers.

Assessments of the effectiveness of the federal labeling program have been limited. The FTC began an evaluation of the labeling program in 1982, but it had not been completed as of 1985. A few studies show that energy efficiency labels have little impact on consumer purchase decisions (McNeill and Wilkie 1979; Anderson and Claxton 1981). For example, a

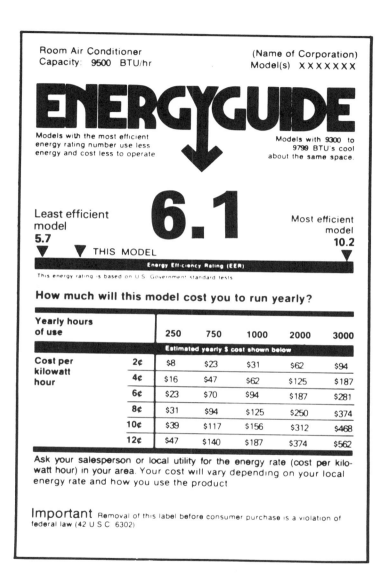

Fig. 9.1. Example of the FTC energy cost label required for room air conditioners.

study of energy labels on refrigerators in Canada showed that labels did not affect their selection; however, labels might increase the pace of efficiency improvements by manufacturers (Anderson and Claxton 1982). Clearly, if labels are to have an impact, efforts must be made to educate consumers about life-cycle costs (Anderson and Claxton 1982).

Residential Conservation Service (RCS)

The RCS, administered by states, requires a local electric or gas utility to provide onsite home energy audits to customers upon request (Walker, Rauh, and Griffin 1985). Information collected during the audit is used to make recommendations on ways and costs of reducing home energy use. The resident is also supplied with additional information on how to implement these recommendations (e.g., lists of contractors who can install recommended measures and names of lending institutions that can finance the work).

By the end of 1983, 40 states had operating programs. From May 1981 to April 1984, 3 million audits were performed at a cost of $360 million, giving an average audit cost of $120 (Department of Energy 1984a). Despite several years of program operation nationwide, little is known about the benefits and costs of RCS or about the key features of the program that contribute to its success or failure (Hirst 1984 and 1985).

Participation in the RCS has been far lower than expected. Audits have been completed for 2% of eligible households per year, or 5.6% of the households offered an audit (Department of Energy 1984a). Some states, such as Michigan and Washington, have experienced much greater response. The low participation rates in general are caused in some cases by ineffective utility marketing, limited local potential for cost-effective conservation, and the absence of a national priority for residential conservation activities (Hirst 1984; Kushler, Witte, and Crandall 1984).

Energy savings resulting from RCS audits also have been lower than expected. Audit programs have resulted in average energy savings of 3 to 5% beyond that attained by nonparticipants. Some evidence suggests additional savings may occur in the second or third year after an RCS audit, as participants gradually implement more recommended measures. For example, the Michigan RCS program found that net savings increased slightly, from 4% to 5%, by the end of the second year.

The overall impact of the RCS program, however, is small. By applying current RCS participation rates and energy savings over a five-

---

The Michigan RCS Experience:
Utility Home Energy Audits Can Work

In the first three years of Michigan's RCS program (launched in mid-1981), 15% of eligible households were audited. This percentage is more than twice the national average participation rate and is attributed to strong support and cooperation by state agencies and utilities (Kushler, Witte, and Crandall 1984). For example, the state PSC quickly established that utilities could recover all costs and the state energy agency thoroughly evaluated all aspects of the program.

Audited households reported taking more conservation measures than unaudited households. Moreover, a study of energy consumption showed that gas consumption dropped an average of 13% two years after the audit in audited homes, 5% more than the average reduction in unaudited homes. Economic analysis shows that the audits and resulting conservation actions are cost-effective from the perspectives of the participant as well as the utility.

Those responsible for planning and implementing the Michigan RCS program are looking towards the future as well as examining the past. They have considerable interest in finding ways to further expand participation, increase consumer response once an audit is provided, and improve audit effectiveness (Kushler, Witte and Crandall 1984). Besides conducting surveys of participants and nonparticipants, the state and utilities are planning to test new, innovative audit schemes, including the installation of a variety of low-cost conservation measures at the time of the audit.

---

year RCS program period, a reduction of only 0.2% in total residential energy use would result. Still, states like Michigan demonstrate that effective RCS programs can be developed and implemented.

Many ideas have been suggested for improving the RCS program (or utility audit programs in general). These ideas include lowering costs by reducing prescriptive requirements and by allowing utilities to employ only those features found to be most useful; allowing the direct provision of conservation measures and products during the audit to ensure that some minimum savings will actually be achieved; and marketing the program to a wider range of customers. Operating the program through an independent nonprofit entity (as is done in Massachusetts and other states) and combining audits with contracting, financing, and inspection (so-called one-stop retrofit service) are also advocated (Sawyer 1985).

STATE PROGRAMS

State Energy Conservation Program

States have supported a wide variety of energy information activities as part of SECP. These activities include brochures, billboards, television and radio commercials, hotlines, and contests. A recent review of evaluations of information provision estimated that information of various sorts leads to a typical energy savings among residential consumers of 5% and that feedback on personal energy use has the potential to achieve residential energy savings of 10%. However, the validity and reliability of many of the information studies are uncertain (Collins et al. 1985).

Energy Extension Service

The Energy Extension Service (EES) is a state-implemented information service funded by the federal government. The program's purpose was to develop a decentralized information system to serve as a local source of information on energy use and efficiency. The program's target audience is small-scale energy users: households, small business, farmers, and local governments.

A survey of state energy officials showed that program administrators believe onsite workshops, auditor training, and carefully targeted information campaigns have been the most effective elements of the EES (Sawyer 1985). General information dissemination is viewed as being least effective. However, attributing a certain level of energy savings to any of these activities is very difficult (Sawyer 1985).

Support of the Private Sector

State information activities can support private sector energy conservation initiatives. Many states cosponsor energy fairs and exhibitions with private organizations to attract more participants than either sponsor could provide on their own. California, Pennsylvania, Maryland, and the Bonneville Power Administration have sponsored seminars and workshops to build credibility for "third-party financing" of efficiency investments and to explain the service. California has also provided guidebooks and model documents for local governments to use in soliciting and evaluating energy service contracts and third-party financing arrangements.

Training and information programs also have been developed to complement regulatory mandates. A review of California's building efficiency standards singles out adequate training of builders and building inspectors as one of several features critical to success (Wicks 1984). The

California experience also proved that good design manuals, periodic newsletters to interpret the regulations and explain compliance options, and a convenient source for immediate interpretation (e.g., hotlines for local code officials) were essential for effectively implementing regulatory programs.

Several studies of state information programs (Collins et al. 1985; Sawyer 1984) found that information is most effective when it:

- Comes from or through a credible, familiar source (e.g., friends, family, neighbors, or community institutions);

- Is provided onsite (e.g., at the home or building where the energy is used);

- Is personalized for the recipient;

- Is presented in clear, concise, simple language;

- Contains specific, concrete, immediately relevant recommendations;

- Directly increases people's knowledge of energy conservation; and/or

- Conveys periodic feedback on the recipient's energy use compared to a previous month or to a community norm.

Unfortunately, many information programs fall far short of these ideals (Stern and Aronson 1984).

Several issues have been identified as critical for increasing the effectiveness of SECP and EES information programs (Cornwall 1984). First, further analysis of which program and organizational attributes are associated with program success is needed. Second, more organized experimentation should be undertaken with different program elements and operations. Third, having up-to-date technical information, information exchange, and better management skills is important for those conducting information programs. Finally, reinforcement activities (e.g., client follow-up or obtaining support from local groups) are critical to improving the effectiveness of information programs.

COMMUNITY PROGRAMS

Both local governments and community organizations have offered energy information, education, and other outreach services for many years. Many of these programs have been financed with SECP and EES funds. Others have been sponsored by existing community organizations whose primary activities include housing services, environmental education, or low-income support services.

Information programs themselves do not directly result in quantifiable energy savings. Therefore, many communities have added hands-on training, cooperative buying, local building regulations, and other "active" elements to their information programs. Other communities found they lacked the desire or resources to do more, and curtailed their energy information programs.

HOME ENERGY RATING SYSTEMS (HERS)

Home energy rating systems have been developed by a variety of federal, state, and local agencies to allow homebuyers and financial institutions to assess the relative energy efficiency and operating costs for new and existing residences. The rating systems are intended to offer for homes the equivalent of the automobile fuel economy ratings or appliance efficiency labels (Fig. 9.2). The ratings may be prescriptive (based on a point system), calculational (based on building and occupant characteristics), or empirical (based on energy bill information). The decision to use these systems is usually made at the local level.

HERS can be used by builders, buyers, real estate agents, mortgage lending institutions, appraisers, and others. Buyers can use the ratings to select a home with lower energy operating costs. Builders and realtors can market energy efficient homes based on the rating. Mortgage lenders can rely on ratings to estimate operating costs and can thereby take energy savings into consideration in qualifying a mortgage loan applicant. Appraisers can use the rating to establish the added home value resulting from the conservation features. The HERS programs that involve all of these parties appear to be most successful.

Unresolved issues pertaining to HERS (Energy Conservation Digest 1984) include the need to increase the reliability of ratings, to further validate rating systems, and to investigate consumer response. In particular, field monitoring to compare the ratings with actual energy use in occupied homes is needed (Rosenfeld and Schuck 1984). Also, the competency and liability of persons performing home energy ratings, the lack of criteria for choosing among existing rating systems, and the use of simple pass/fail systems (which discourage high levels of energy efficiency) are areas of concern.

Perhaps the most outstanding needs are sponsor support and promotion to bring HERS into widespread use by lenders, homebuyers, and realtors. Home energy ratings could be tied to other programs, such as utility audits or normal real estate appraisals. The acceptance of HERS should increase if programs incorporate research on motivating behavior, such as

HOME ENERGY RATING

Prepared for: Name Joseph Smith
Address 33 Centre Street,
Auditor Signature Cristin
Date 10/1/82 Audit # 701100

HOME NOW	IMPROVED HOME
Heating Efficiency Score	**Heating Efficiency Score**
0 1 2 3 4 (5) 6 7 8 9 10	0 1 2 3 4 5 6 (7) 8 9 10
Worst (No energy features) — **Best** (No heating bill)	**Worst** (No energy features) — **Best** (No heating bill)
$ approx $1450	$ approx $1050
Estimated Annual Heating Cost	**Estimated Annual Heating Cost**

Rated Features now in the home:	ENERGY FEATURES	Add these Features for improved Rating:
very leaky — leaky — (moderately tight) — tight — very tight — N/A	INFILTRATION LEVEL	Improve to "moderately tight" (add caulking & weatherstripping)
R-0 — R-11 — (R-19) — R-30 — R-38 — N/A	CEILING INSULATION	Add (R-11), R-19, R-30, R-38 To get a total of (R-30), R-38
R-0 — R-7 — (R-11) — R-19 — R-19+ — N/A	WALL INSULATION	Add R-11
(R-0) — R-7 — R-11 — R-19 — R-19+ — N/A	FLOOR INSULATION	(Add R-19)
single pane — single with drapes — double pane — (double with drapes) — triple — N/A	WINDOW TREATMENT	Add storm windows or replace sash to get a double layer on each window
absent — installed — (N/A)	NEW HEATING SYSTEM	Install
(absent) — installed — N/A	FLAME RETENTION BURNER	(Install)
(absent) — installed — N/A	PIPE INSULATION	(Install)
absent — installed — (N/A)	DUCT INSULATION	Install
		Estimated Improvement Costs: $ 1525
Features in the home, but not rated:		**Add these important Features too. (Not rated)**
absent — installed — (N/A)	WINDOW INSULATION	Install
absent — (installed) — N/A	CLOCK THERMOSTAT	Install
absent — (installed) — N/A	WATER HEATER INSULATION	Install
absent — installed — (N/A)	SOLAR WATER HEATER	Install
absent — installed — (N/A)	VENT DAMPER	Install
absent — installed — (N/A)	ELECTRONIC IGNITION	Install
absent — installed — (N/A)	NEW COOLING SYSTEM	Install
absent — installed — (N/A)	SOLAR POOL HEATER	Install
absent — installed — (N/A)	HEAT GAIN RETARDANT	Install
	OTHER	

Home Now	Note	Improved Home
Btu's/Square foot × degree days 10	**Note:** The Predicted Annual Heating Cost for the house is based on the following assumptions. Electricity $/kwh – Heating Oil $/gal 1.25 Natural Gas $/ccf –	Btu's/Square foot × degree days 8

Disclaimer: Although every effort has been made to provide accurate information on this form, neither this form nor any entries made on it constitutes any warranty, express or implied, as to, without implied limitation, the presence or lack of energy features in the house, the heating fuel used in the house and its cost, or the actual annual heating energy consumption of the house.

Fig. 9.2. The Massachusetts Home Energy Rating form.

The Uniform Energy Rating System:
An Exemplary Home Energy Rating Program

The Uniform Energy Rating System (UERS) was developed by the Western Resources Institute in Seattle in conjunction with lenders, appraisers, builders, and the Federal Government (McCarty and Willner 1985; Frahm 1984). Appraisers trained by the Institute rate homes on a scale of 0 to 100 based on all aspects of energy efficiency. The overall efficiency rating is then used to estimate annual energy use and annual energy cost.

As of mid-1985, 12 major banks in the Puget Sound region use UERS before considering a mortgage or home improvement loan. These banks increase their maximum loan offer by up to 10% for homes rated efficient or very efficient. In addition, the Institute actively promotes the program, and utilities have also expressed an interest in adopting the ratings.

The UERS rating is based on detailed descriptive information on the characteristics of a home. Comparison of the ratings and actual energy consumption shows that UERS is generally within 15% of actual usage. As of 1985, 1000 homes per year were receiving ratings, and the program was still expanding. The success of the program is attributed to support from all relevant sectors: banks, realtors, appraisers, builders, utilities, and consumers.

---

emphasizing monetary benefits and interpersonal interaction rather than general information provided through mass media (McCarty and Willner 1985). Finally, efforts should be made to extend energy ratings to rental housing and commercial buildings (Rosenfeld and Schuck 1984).

C. FINANCIAL INCENTIVE AND GRANT PROGRAMS

Economic incentive and grant programs have been developed (primarily by the federal government) to motivate greater investment in energy efficiency. As shown in Table 9.2, federal expenditures for tax credits, low-income weatherization, and grants to states and institutional energy users ranged from $700 to 850 million per year between 1978 and 1983. Several states also offer tax credits for conservation investments in buildings.

FEDERAL INCENTIVE PROGRAMS

The federal government offers four primary forms of economic incentives to stimulate energy conservation actions. These are residential energy conservation tax credits, which encourage residents to install conservation measures in their dwellings; the Institutional Conservation Program, which pays for audits and half the conservation investments in eligible school and hospital buildings; the Weatherization Assistance Program, which funds installation of conservation measures in the homes of low-income households; and the Solar Energy and Energy Conservation Bank, which provides grants and loan subsidies. Only the last program includes apartment building owners or the commercial sector outside of schools and hospitals to a significant extent, and owners of single family housing still receive the bulk of the funds in this program.

Tax Credits

The residential energy conservation tax credit amounts to 15% of the conservation measure cost, with the credit not to exceed $300. The occupant (owner or tenant) of any dwelling is eligible. The conservation tax credit, which began in 1978, expired at the end of 1985. Between 1978 and 1983, 24 million households claimed a residential energy tax credit (no correction for repeat-year applications), and the lost revenue to the federal treasury totaled $2.3 billion (Fig. 9.3).

Although several studies have been carried out to determine the extent to which tax incentives stimulate conservation, the results are inconclusive (General Accounting Office 1982a; Hirst et al. 1983; Energy Information Administration 1985). Some evidence exists that the amount of the credit was too small to significantly affect behavior and that investment decisions are based more on expected savings and rising energy prices than on initial capital cost. A larger credit might substantially lower the cost of saved energy to the federal government (Department of Energy 1984b). Unfortunately, many of the assessments of the federal tax credit are based on self-reports regarding consumer behavior, and their reliability is difficult to judge (Hirst et al. 1983).

The residential tax credit has had a negligible impact on rental housing, and participants are skewed towards upper-income groups (Department of Energy 1984b). If the tax credit is continued, it would be useful to redress this equity problem. Proposals to extend the federal tax credit, such as the bill introduced in the 98th Congress (H.R. 6244), include an eligibility ceiling on gross income (Congressional Record 1983). The proposed tax credit also raises the credit percentage to increase the incremental investment in energy conservation.

Institutional Conservation Program

The Institutional Conservation Program (ICP) is funded by the federal government and administered by the states. It provides funds that allow nonprofit institutions to assess energy savings potential and to undertake

Table 9.2. **Federal expenditures for energy conservation incentives and grants programs (million $)**

Program	Year								
	1976	1977	1978	1979	1980	1981[a]	1982	1983	1984
Treasury									
Residential Conservation Tax Credit			558	435	415	364	292	243	NA[b]
DOE									
Institutional Conservation (Schools/Hospitals/Local Government)				107	162	150	48	98[c]	48
Weatherization Assistance	44[d]	132[d]	130[d]	199	199	175	143	245[e]	190
State Energy Conservation	5	35	72	58	48	48	24	24	24
Energy Extension Service		8	8	15	25	22	10	10	10
DOE Subtotal	49	175	210	379	434	397	225	377	272
HHS									
Low-Income Home Energy Assistance (Weatherization Component)							136[f]	195[f]	196[f]

Table 9.2. (continued)

Program	Year								
	1976	1977	1978	1979	1980	1981[a]	1982	1983	1984
HUD									
Solar Energy and Energy Conservation Bank							22	20	25
Grand Total	49	175	768	814	849	761	675	835	NA[b]

[a]After rescissions and deferrals adopted by Congress in FY 1981.
[b]Not available.
[c]Includes $50 million from the 1983 Jobs Act.
[d]In Fiscal Years 1974 to 1978, weatherization was funded in whole or in part through the Community Services Administration.
[e]Includes $100 million from the 1983 Jobs Act.
[f]Beginning in FY 1981, Congress gave states discretion to spend up to 15% of their LIHEAP funds on weatherization. These numbers are Department of Health and Human Services estimates of the total used by all states for this purpose.
Source: Subcommittee on Energy Conservation and Power, U.S. House of Representatives.

energy related capital improvements to their facilities. Nonprofit institutions tend to be less responsive to energy saving opportunities because of budgetary constraints and because they cannot take advantage of tax credits. Hospital energy costs are passed through to patients and third-party reimbursers.

The ICP authorized states to make available on a competitive basis to schools, hospitals, local governments, and public-care facilities (e.g., convalescent homes) funds to conduct detailed energy audits of their facilities. Additional grant support of capital retrofit expenditures is available to qualifying schools and hospitals only. Funding in recent years has *not* included sufficient money for energy audits of local government or public-care facilities (Bakken 1984).

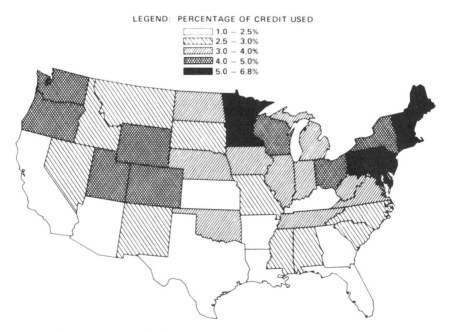

LEGEND: PERCENTAGE OF CREDIT USED

- 1.0 – 2.5%
- 2.5 – 3.0%
- 3.0 – 4.0%
- 4.0 – 5.0%
- 5.0 – 6.8%

Fig. 9.3. Percentage of the potential federal residential tax credit used between 1978 and 1980, by state (Hirst, Goeltz, and Manning 1982). Note that only a small fraction of the available tax credit was claimed during this three-year period. Also, there was tremendous variation among states in applications for the tax credit during this time.

Assessments of the ICP program have shown that the audit phase of the program results in little energy savings (Department of Energy 1983). However, the energy savings from the equipment grant phase of this program amount to 13% of preprogram energy use. Assuming a 10-year equipment life, the average cost of saved energy for these investments is estimated to be about $1.40/MBtu (Department of Energy 1983), far below the average fuel prices paid by these institutions. Total capital costs are typically paid back in two to three years.

Although the present Administration has tried to terminate ICP, state officials are enthusiastic about it and estimate that fewer than 25% of the capital improvements would have been made without federal assistance (Sawyer 1985). These officials consider the program to be twice as effective as the EES, Weatherization Assistance, and RCS programs and 50% more effective than the SECP program in terms of relative savings per dollar of government expenditure. The greatest shortcomings of ICP are the limited funding and the exclusion of local government and residential-care facilities from eligibility for capital grants (Sawyer 1985).

Low-Income Weatherization

Low-income households have a serious energy problem: their homes tend to be older energy wasters, and they lack the money to make improvements. The Weatherization Assistance Program (WAP) provides funds to states to weatherize low-income households' dwellings at no charge to the resident.

The WAP program in seven years has weatherized approximately 1.5 million households of the 13.1 million eligible (Sawyer 1984). At this rate, it will take more than 50 years to reach all those that are eligible. Furthermore, the WAP program, administered by state and local social service agencies, has been hampered by competing social goals (job training and reemployment objectives), bureaucratic requirements, uncertain energy savings, and an overemphasis on high-cost conservation treatments (Sawyer 1984; General Accounting Office 1981).

Research projects and demonstration programs have shown that properly carried out retrofits of low-income housing can save substantial energy at reasonable costs. For example, the Community Services Administration and the National Bureau of Standards conducted a demonstration program involving building-envelope and/or heating system modifications in 142 homes (using local weatherization crews). This effort resulted in an average reduction in space heating demand of 31% with an average first cost of $1,610 and an average simple payback period of eight years (based on 1979 energy prices) (Crenshaw and Clark 1982).

Some WAP experience is consistent with this demonstration program. Data on retrofits of 938 homes in 27 WAP programs show an average space heat savings of 24%, average first cost of $1,580, and an average simple payback period of 11 years (Goldman 1984).

Other survey results are less encouraging. A sample of more than 1700 low-income homes weatherized in 1981 show an average space heating fuel savings of only 13 to 14%; 23% of the households sampled actually use *more* space heating fuel in the post-weatherization period (Peabody 1984).

A major evaluation of Wisconsin's low-income weatherization program (which received $13 million in federal funds in 1984) showed a typical heating energy savings of only 6 to 10% at an average total cost (including materials, labor, and program support and administration) of $2250 per household (Hewitt et al. 1984). Several recommendations were made for increasing the effectiveness of this program. Better monitoring of local programs with respect to quality of work and selection of weatherization measures was called for. Also, a new ranked list of weatherization activities was developed featuring the use of blower-door tests for detecting infiltration losses, routine sidewall insulation, and routine furnace retrofit or replacement. This example demonstrates that a full-scale residential conservation program can be much less successful than research or pilot projects, and that careful program evaluations are important.

Several efforts have been made to use lower-cost conservation packages to improve the cost-effectiveness of WAP programs and to assist more eligible households. The City of Cleveland operates such a program, described later in this chapter. Sun Power Consumer Association in Colorado operates a furnace efficiency improvement program at a cost of $150 per unit (Proctor 1984). The measured savings averaged 12% of an annual heating bill of $724, yielding an average payback of less than two years.

Another funding source for low-income weatherization is the Low Income Home Energy Assistance Program (LIHEAP, Fig. 9.4). That program is administered by the U.S. Department of Health and Human Services, which allocates funds to states to assist households that cannot pay their utility bills. In 1981 to 1983, it allocated $1.9 billion per year. Fifteen percent of these funds may be set aside to finance low-cost, energy-conserving improvements. However, actual set-asides have amounted to just over half the potential. Only 20% of the set-asides have gone to innovative conservation financing; most have been added to existing WAP services (Ferrey 1984).

Several concepts have been proposed to increase the leverage of low-income program funding. Ferrey (1984) proposed using multiple sources of funds (e.g., LIHEAP, Solar and Conservation Bank, WAP, oil overcharge refunds, and housing program funds) to leverage private financing of conservation improvements and to bridge current program gaps. Rather than using limited funds to finance 100% of conservation measure costs, it may be possible to obtain matching funds or to provide loan-interest subsidies, loan down payments, or loan security.

Solar Energy and Energy Conservation Bank

The SEECB was established by Congress in 1980 to provide financial assistance for cost-effective solar and conservation measures in low- and moderate-income housing and commercial buildings with nonprofit owners or tenants. Federal funds are distributed to states, which then provide loan subsidies and matching grants through financial institutions within the state.

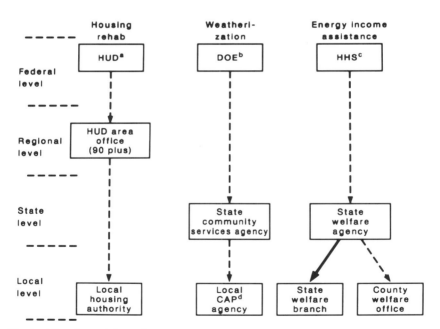

[a]Dept. of Housing and Urban Development
[b]Dept. of Energy
[c]Dept. of Health and Human Services
[d]Community Action Program

Fig. 9.4. Many government agencies at the federal, regional, state, and local levels are involved with energy-efficiency improvements in low-income housing.

The SEECB got off to a shaky start, and final regulations governing operation of the SEECB were not issued until 1984.

As of 1985, 38 states had received more than $18 million with an additional $42 million obligated (Preysner 1985). The vast majority of the funds were for conservation projects in lower-income single-family housing, although conservation projects in multifamily housing are becoming more popular.

Nearly two-thirds of the currently obligated funds are for loan subsidies, where the average subsidy is equal to 38% of the measure cost (Housing and Urban Development 1984). It is estimated that the conservation measures being installed under the program have an average payback period of 3.7 years (Housing and Urban Development 1984).

While it is too early to pass judgement on the SEECB, some positive developments have occurred. First, a number of states are using creative techniques to promote the program, including working with existing neighborhood or utility programs and block-based marketing (Ferrey 1984). Second, some innovative schemes are underway to leverage SEECB funds, increase the effective subsidy for low-income persons, and involve owners of multifamily housing to a greater extent (Ferrey 1984; HUD 1984).

STATE PROGRAMS

Many states offer incentive programs to supplement federal incentives. States with their own conservation tax credits as of 1983 include Arizona, California, Colorado, Oregon, and Rhode Island (Randolph 1984). California spent $38 million in 1983 on their 40% conservation tax credit (California Energy Commission 1983c). However, it is estimated that 43% of this amount was returned to the state through other tax revenues, both directly from the conservation activity and indirectly through the increase in disposable income resulting from lower energy bills (California Energy Commission 1983c).

Some states instituted their own loan programs. For example, California established a $20 million loan pool to help schools, hospitals, local governments, public-care facilities, and special districts undertake energy audits and conservation measures. The Minneapolis Community Development Agency issued revenue bonds to subsidize financing (10% interest, 10-year loans in 1982) for efficiency improvements in multifamily housing. The state of Minnesota also encourages private-sector loans to rental housing through a loan insurance program for lenders (Bleviss and

Gravitz 1984). New York's Energy Investment Loan Program is targeted at small and medium-size manufacturing firms, multifamily housing owners, and not-for-profit organizations. Other states with loan programs include Alaska, Oregon, Maine, and Maryland (Randolph 1984).

States also are using seed money to help local government and private organizations launch self-sustaining energy conservation activities. For example, the North Carolina Alternative Energy Corporation's Local Energy Officer program guarantees the first two years of a local government's energy manager's salary in the event that not enough energy savings are produced to cover staff costs. North Carolina expects to pay out in very few cases and appears to have established a successful program (Gee 1984). Similarly, Missouri provides grants and technical assistance to support full-time energy management offices in local governments for the first year. Program costs for subsequent years are then funded out of previous-year savings (Collins et al. 1985). Few (if any) evaluations of the incremental energy savings and cost effectiveness of state loan and incentive programs have been performed (Randolph 1984).

LOCAL PROGRAMS

We are not aware of incentive or grant programs offered by local governments or community groups. Financial subsidies have been incorporated into some of the broader energy service programs described later in this chapter.

Some communities have combined multiple sources of funding and expertise into an integrated low-income assistance program. In 1983, Cleveland was faced with a situation where 59,000 households qualified for utility bill assistance under the LIHEAP program while only 865 homes could be weatherized through WAP. To increase energy conservation services for the other 58,000 families, Cleveland brought city agencies, neighborhood organizations, community development corporations, and a university together to deliver integrated audit, weatherization, retrofit, and loan programs (Krumholtz and McDermott 1984). Funding comes from three foundations, a local utility, a major corporation, and the city. The program is piggybacking conservation services onto established housing-rehabilitation, neighborhood-outreach, and fuel-bill-assistance services (Mihaly 1984). The weatherization program is also being used for job training and human development. The program installs low-cost storm windows, caulking, and door weatherstripping at an average cost of only $200 per home (Mihaly 1984), with an estimated 15 to 25% reduction in natural gas use.

D. REGULATORY PROGRAMS

FEDERAL PROGRAMS

Legislation passed in the 1970s (Table 9.1) ordered DOE to promulgate appliance efficiency standards and Building Energy Performance Standards (BEPS). However, the Federal Government during the 1980s has opposed federal promotion of energy conservation through standards setting and so far has not adopted national appliance efficiency standards or mandatory BEPS. Nevertheless, significant benefits could be derived from federal standards.

Appliance Standards

In the case of appliances, national standards could eliminate the problem of manufacturers' confronting a host of varying state regulations. An analysis of appliance standards conducted for DOE estimated that standards could reduce residential energy demand by 1.1 Quads per year or 6.4% by the year 2005 and save consumers $10 to 16 billion (net present value benefit over a 20-year period) (Levine et al. 1984). DOE, however, declined to issue standards, arguing that standards would result in insignificant energy savings, would have an adverse impact on manufacturers and lower-income persons, and would reduce product utility.

DOE's ruling was criticized for using outdated data, unsubstantiated and optimistic assumptions about the behavior of the marketplace in the absence of standards, and misinterpretation of the legislative requirements (General Accounting Office 1982b; Geller 1983; Geller and Miller 1984; Rollin and Beyea 1985). In July 1985, the Court of Appeals overturned DOE's ruling and provided clear instructions to DOE for issuing substantive standards (U.S. Court of Appeals 1985).

BEPS

Strong energy standards for new buildings were called for in the 1976 ECPA legislation. The reduction in energy use for space heating and cooling in new single-family housing resulting from the BEPS standards was originally estimated to be roughly one-third to one-half compared to the energy use resulting from conventional practice (Department of Energy 1979). However, designers' and builders' organizations strongly opposed the BEPS program, and it was subsequently converted to voluntary guidelines and design tools for designers and builders.

As part of the "new" BEPS program, a set of simple "slide rules" have been developed that a designer or builder can use to determine the energy costs associated with different envelope and space conditioning equipment combinations (Ritschard et al. 1984). The slide rules should also allow builders, architects, and local building code officials to compare the energy performance of housing designs with the voluntary BEPS.

An important but largely unresolved issue concerns the need for direct government intervention in the building design process. Can existing state standards along with information and education programs at the national level lead to new buildings with a high degree of energy efficiency? This question can be answered objectively only if data are collected to see if builders are producing energy efficient buildings on their own without regulatory impetus.

Other Construction Standards

The Department of Housing and Urban Development has Minimum Property Standards (MPS) that include efficiency provisions for federally financed housing. The MPS requirements result in homes with up to 35% lower heating loads than those meeting ASHRAE-based standards (Kelly 1984). New homes must satisfy the MPS requirements to qualify for FHA or VA loans, and about 40% of new single-family housing units met the MPS requirements in recent years (Kelly 1984).

Federal Manufactured Home Construction and Safety Standards (MHCSS) establish minimum quality requirements for manufactured housing. These standards are designed and administered by the Department of Housing and Urban Development (HUD). Unfortunately, the current MHCCS are very weak. HUD proposed new standards in 1983, but the Office of Management and Budget blocked their adoption. The proposed new standards for manufactured housing still fell far short of the MPS standards for site built homes and life-cycle cost minimum levels (Woodham 1984).

STATE REGULATORY PROGRAMS

Building Codes

Many states have adopted thermal efficiency standards for new building construction based on the model standards developed by the American Society of Heating, Refrigerating, and Air Conditioning Engineers (ASHRAE 90A-75 and 90A-80). It appears, however, that the average efficiency of new homes in most states significantly exceeds these standards (McCold 1984). It is therefore unclear what effect standards have had on improving building thermal performance.

The implementation and enforcement of state and local building codes are highly variable. States differ substantially in terms of the types of buildings covered (residential, commercial, and government-owned); strength and enforcement of codes (guidelines, voluntary standards, and mandatory standards); and universality of coverage (local option, state standards implemented by local governments, and state standards implemented by state agencies) (Collins et al. 1985).

A few states, such as California, New York, Minnesota, South Dakota, and Wisconsin, have adopted stricter building standards. California, for example, began upgrading its commercial building standards in 1984. These innovative standards are designed to achieve a high level of energy efficiency by encouraging passive solar design and state-of-the-art technologies through both prescriptive and performance-based requirements (California Energy Commission 1983a). The Northwest Power Planning Council proposed stringent standards for new homes and commercial buildings with electric heating in the Pacific Northwest (Northwest Power Planning Council 1983).

A major issue related to the design of building codes is the relative merits and desirability of performance-based and prescriptive standards. Performance standards give designers, builders, and retrofitters maximum flexibility in achieving efficiency goals. However, performance-based standards are more difficult to enforce than standards that prescribe certain levels of thermal integrity in different building components.

Some states are experimenting with a combination of performance-based and prescriptive standards. For example, Florida adopted a model energy efficiency code for new construction in 1982 that is based on a point system (Florida Department of Community Affairs 1982). It permits tradeoffs between the efficiency of the building envelope and that of the space conditioning equipment. In 1984, South Dakota adopted a new code requiring homes financed through the state Housing Development Authority either to meet a strict overall performance standard or to include certain specified measures (Nisson 1984).

The energy saving effects of new building thermal performance standards are likely to depend on the training and education of builders and local code officials, the degree to which these officials enforce the code (through, for example, onsite inspections rather than review of builders' plans only), the length of time the code has been in effect, the complexity of the code, and the frequency with which the code is modified. Unfortunately, little evidence is available to document the effect these factors have on building energy performance.

Two states, Minnesota and Wisconsin, have adopted regulations on the energy efficiency of existing buildings including standards for rental housing. These steps were taken because rental-housing owners have insufficient motivation to invest in conservation measures. However, because of problems in prescribing multifamily-housing efficiency improvements as well as in inspection and enforcement, the Minnesota code (enacted in 1978) has had little impact so far (Bleviss and Gravitz 1984). The Wisconsin code, which went into effect in early 1984, is too recent to evaluate, but early indications are encouraging (Bleviss and Gravitz 1984). Both rental housing codes primarily focus on the building shell with little attention to HVAC systems.

Appliance Standards

Several states have enacted minimum efficiency standards for appliances sold within their borders (Randolph 1984). California adopted standards for a wide range of products during the 1970s (California Energy Commission 1983b). Other states, such as New York and Kansas, have minimum efficiency standards for air conditioners, and nearly all states have adopted new building codes that include equipment efficiency requirements based on ASHRAE recommendations (Geller 1983).

Appliance Standards in California:
A State Takes the Lead in Efficiency Regulation

California adopted minimum-efficiency standards for refrigerators, freezers, and air conditioners in 1976 and for water heaters, furnaces, heat pumps, and other gas-fired products subsequently (CEC 1983b). The program operates with minimal inconvenience to dealers and consumers and a high level of compliance.

The California Energy Commission estimates that appliance standards will reduce peak power needs by 1750 MW and save consumers more than $600 million by 1987 (CEC 1983b). This makes the standards the most effective conservation program adopted by the Commission. The savings estimates do not, however, account for the natural adoption of efficient equipment in the absence of standards.

By 1983, California's standards for refrigerators and freezers had become outdated (i.e., all models being manufactured qualified for sale in California). The Energy Commission proceeded to analyze, propose, and adopt new standards for these products as well as for central air conditioners and heat pumps in 1984 and 1985. The ambitious new standards for refrigerators and freezers require by the early 1990s efficiency levels beyond those currently available in the U.S.

In general, state energy policy makers consider efficiency regulations one of their most effective activities (Sawyer 1984). Interest in appliance standards among states has grown since the federal government has failed to adopt national standards. Maine adopted new comprehensive appliance standards in 1985, and other states have standards under consideration.

Experience has shown that state appliance efficiency standards can become outdated quickly. To increase energy savings, New York raised its central air-conditioner standard to a stringent 9.5 seasonal EER (SEER) level in 1984. (For comparison, the typical central air-conditioning system sold in 1984 had an SEER rating of 8.6, while highly efficient units have SEER ratings of more than 12.) California tightened its standards on some products in 1984.

Several uncertainties are associated with appliance efficiency standards. First, how appliance manufacturers respond to individual state standards is not clearly understood. Goldstein (1983) argued that the 1978 refrigerator standard in California affected product offerings nationwide. However, manufacturers claim that the general improvement in the efficiency of refrigerators since the late 1970s was caused by marketplace pressures and not by federal or state regulatory activities (Association of Home Appliance Manufacturers 1982).

Second, the impact that standards have on new-product efficiencies beyond changes that naturally occur in the marketplace is difficult to estimate. More work is needed to understand purchaser behavior and the impacts of standards (as well as of alternative programs) in different market segments (e.g., new-home buyers, low-income consumers, and emergency replacements). In addition, the potential impacts of appliance standards on electric utilities (such as reductions in peak load and the provision of "assured savings" for inclusion in a utility's resource plan) have not been well established.

REGULATIONS AT THE LOCAL LEVEL

Local governments in some areas have adopted energy retrofit ordinances requiring a certain level of conservation to be achieved in existing residences. The local ordinances may be targeted at all housing, rental housing only, single-family homes only, or multifamily housing only. The degree of energy efficiency to be achieved varies, as do the requirements for compliance. Some ordinances require compliance before a home can be sold, others require the home buyer to install minimum levels of conservation, and at least one sets a date by which all buildings must be in compliance regardless of whether a sale occurs.

E. ENERGY SERVICE PROGRAMS

Most of the programs described above are single-purpose programs intended to deliver information or incentives or to mandate energy efficiency. In recent years, program operators have begun to modify and expand their efforts to increase program effectiveness. We are now seeing the emergence of multifaceted energy service organizations that deliver information, analysis, products, installations, financing, and/or ongoing maintenance. Their common emphasis is on results.

In Minneapolis, neighborhood-based workshops are generating a high level of response. Integrated conservation delivery programs involving education, audits, financing, grants, contracting, and follow-up are also being undertaken in Massachusetts (Cowell and Rebitser 1984).

The Minneapolis workshops teach residents how to do no-cost or low-cost efficiency improvements; participants also receive some weatherization materials and an RCS audit (Brummitt 1984). Participants are often encouraged to make major weatherization investments and are offered low-interest financing to do so.

By employing group motivational techniques, participation averages 30 to 40% of the households in blocks served, approximately 10 times the participation rate common to RCS programs. Between 1981 and mid-1985, 28,000 households participated in the workshops. The entire program is delivered at a cost of about $80 per household, including a $30 audit.

An evaluation of the program showed that participants making the no-cost and low-cost improvements reduce energy use by more than 7%, compared to 2 to 3% for nonparticipants (Brummitt 1984). As of 1985, about 75% of the homeowners receiving recommendations for major weatherization work actually go ahead to some extent (Esposito 1985).

In Santa Monica, California, a program started in 1984 offers "generic" audits on a door-to-door canvassing basis combined with on-the-spot installation of low-cost conservation devices free of charge (Egel 1984). All residents are able to participate: tenants or owners; single-family or multifamily dwellers. Audits are being completed in 30 to 35% of the households approached, with nearly all receiving at least one conservation device.

Government organizations are becoming involved with energy service companies in a number of ways. At the local level, energy service arrangements are emerging as a means to upgrade the thermal integrity of government buildings and facilities (Weedall 1984). At the state level, government agencies are promoting third-party, energy service transactions through technical assistance, demonstration, networking, and regulatory activities. At the federal level, tax incentives stimulate establishment of third-party financing and energy service companies in the private sector.

The government can facilitate the use of energy service arrangements in multifamily housing. This sector is often shunned by private energy service companies (Weedall 1984). For example, New York City has contracted for energy management and conservation financing for 30 to 60 apartment buildings that it owns (Brennan and Zelinski 1984). In Massachusetts, the nonprofit Citizens Conservation Corporation (CCC) provides energy services in lower-income apartment buildings. They enlist tenant cooperation by offering them a share of the savings (Haun 1984). In St. Paul, Minnesota, the Energy Resource Center (ERC) began an energy management service for multifamily rental buildings in 1983 (Griffin et al. 1984). Both the CCC and ERC programs relied on government agencies for seed money and for operating subsidies in their initial years.

The energy services field is complex and growing in sophistication. Maintaining knowledge of technologies and practices and retaining a competent staff require information exchange and possibly the use of outside technical assistance. For a community organization, satisfying these requirements may mean establishing ties to a local university, energy center, information clearinghouse, and/or local utility.

A number of research activities related to third-party financing and energy service companies are needed. Additional projects need to be monitored and documented to increase general awareness and credibility. Additional efforts are needed to develop standard contracts and other support materials. Research is needed on techniques for determining energy savings in the wide range of commercial/institutional building types (Weedall 1984).

Although the energy services concept appears promising, it cannot come into fruition overnight. Private- and public-sector organizations active in this field suggest it will take 5 to 10 years for the energy services industry and supporting mechanisms to fully develop (Weisenmiller 1984). Sustaining progress during this period will depend on a consistent economic and tax environment and expanding information, assistance, and research activities.

F. SUMMARY AND CONCLUSION

Government programs to promote energy conservation in buildings are very broad in scope. Planning for energy efficiency started with an enthusiastic "let's do something for energy efficiency" when little was known about how to achieve this. Thus, early planning efforts were general in nature and lacked well-defined actions or objectives. More recently, regional, state, and local organizations have shown a growing sophistication in developing targeted conservation plans with well-defined program elements. This sophistication could be enhanced by improved information sharing about technologies, program experiences, and planning efforts.

Information programs have been addressed both to the consumer of energy and to those who influence energy use through the design of buildings, the financing of construction projects, and the selection of appliances and equipment. The few systematic studies of consumer information programs generally conclude that these programs have had small direct impacts on energy conservation.

Analyses by psychologists and marketing professionals indicate that many information programs were not designed to serve as a basis for action. The programs overlooked the importance of identifying consumer needs and relating information to the consumer's decision-making process. Such information must present the personal benefits to be obtained, a reason to act now, and the steps needed to be taken to obtain the benefits.

The federal government funds and states often manage various incentive or grant programs for improving energy efficiency in buildings. Federal funds were expended at nearly $800 million per year from 1978 to 1982 with about half this expenditure provided through the residential conservation tax credit.

Unfortunately, how much additional investment in conservation is taking place as a result of these expenditures and how incentives influence energy investment decisions are not known. Several fundamental questions remain unanswered: (1) How do incentives affect decision-making? (2) Are incentives an economic persuader or are they primarily a marketing tool to capture consumer attention? (3) What size and form (e.g., loan vs cash rebate, or low- vs no-interest loans) of incentives are needed to motivate desired conservation actions among different energy users? (4) How can nonenergy impacts, such as health and environmental benefits or employment creation, be accounted for?

Efficiency standards to reduce energy consumption of building envelopes, appliances and other equipment have proven to be a low-cost and effective policy option. The current administration, though, opposes implementation of federal efficiency standards for building envelopes and appliances. States, on the other hand, have embraced a variety of energy efficiency standards. When such standards are adopted, they must be kept up-to-date.

The emerging area of comprehensive energy services combining information, technical assistance, and "delivery" of energy savings through equipment sales, installation, and/or financing holds great promise. Governmental authorities can foster this "energy services industry" by using such services at the local level and by providing demonstrations, technical assistance, economic incentives, and research support at the state and federal levels.

After the burst of enthusiasm for government-sponsored conservation programs in the late 1970s, followed by the budget reductions and challenges of the 1980s, several broad lessons and themes emerge.

The Importance of Program Evaluation

A program that works in one instance can turn out to be a failure when applied elsewhere. Hence, greater emphasis needs to be placed on evaluating programs to understand why certain results were obtained, to identify the essential attributes of successful programs, and to improve poorly performing programs.

The Importance of Market Research and Marketing

The skills of psychologists, market researchers, sales and installation personnel, and product managers/implementators must be applied to conservation programs. Alternative program designs must be tested from the financial, administrative, and technical perspectives. Also, effective marketing channels are needed along with techniques for achieving the desired energy savings in the most cost-effective manner in different market segments.

Role of the Federal Government

Clearly stated federal policies supporting improved energy efficiency are needed to ensure an adequate nationwide focus on the issue and to motivate action at various levels. This is not to say that the federal government must develop specific programs in all cases. On the contrary, evidence exists that a policy umbrella is needed from the national government, bolstered by reasonable levels of funding, technical assistance, and "proof-of-concept" or demonstration programs. National initiatives are clearly important in some areas (e.g., efficiency standards where the unregulated marketplace is not providing cost-effective efficiency investments). In other areas, national efforts should give state and local governments the flexibility to design and carry out their own programs. Likewise, government policies and programs should preserve opportunities for business and professional organizations to initiate energy efficiency strategies consistent with national objectives.

Role of State and Local Authorities

State and local officials can develop programs that are responsive to local needs and capabilities and are the ones best equipped to integrate energy efficiency components within existing nonenergy programs and institutions (Sawyer and Armstrong 1985). The needs for broadly implementing conservation programs at the state and local level include adequate funding and development of techniques for obtaining and using nongovernment funds (e.g., third-party financing). Also, small, decentralized program operators need to be provided up-to-date and reliable information on technologies and program strategies.

Low-Income Housing

Low-income persons generally live in energy-wasteful housing and face crippling energy bills. A concerted effort is needed to both expand and improve low-income weatherization efforts through more effective retrofits (e.g., low-cost measures along with furnace modification or replacement), better program management (e.g., more inspection and quality control), innovative financing schemes (e.g., leveraging funds from public and private sources), better outreach (e.g., neighborhood workshops), and regulatory approaches (e.g., rental housing codes).

Program Efforts in the Commercial Sector

So far, government-sponsored conservation programs have been directed primarily at single-family residential consumers. Although some use has been made of new construction standards, lighting efficiency standards, and grants for nonprofit institutions, the commercial sector deserves greater attention in programmatic efforts. Issues that merit further study in these areas include organizational behavior as it affects energy use; the degree to which information, incentives, etc. must be tailored to the owners of different classes of commercial buildings; and techniques for motivating conservation actions in these building types.

10

UTILITY CONSERVATION PROGRAMS

A. BACKGROUND

We discuss gas and electric utilities separately from government and community organizations for several reasons. First, utilities have frequent (generally monthly) contact with all their customers. Second, utilities have information on their customer's use of fuel: how much is used, when it is used, and how much it costs. Third, utilities are often able to capture benefits of energy efficiency improvements not available to energy users themselves. For example, residential customers who face a fixed price for electricity pay the same regardless of whether consumption occurs at the time of the utility system peak or not. Therefore, the household has no incentive to reduce consumption during times of system peak. However, the utility has considerable incentive to reduce system peak and thereby lower its operating costs. Fourth, utilities already operate many energy conservation and load management programs because of regulatory and public pressure, concern for their customers, and the economics of operating their system. Finally, the utility industry research organizations, Electric Power Research Institute and Gas Research Institute, sponsor substantial energy efficiency R&D programs.

This chapter is divided into two portions. The first, analogous to the earlier discussion of human behavior in Chapter 8, deals with utilities as institutions and with the forces that affect their decisions to operate energy conservation programs. The second portion, analogous to the discussion of government and community programs in Chapter 9, describes utility programs themselves. The programs are discussed in the same manner as in Chapter 9: planning, information, financial incentives, and hybrid services. For obvious reasons, utilities do not operate regulatory programs.

B. UTILITIES AS INSTITUTIONS

RECENT TRENDS IN UTILITY CONSERVATION ACTIVITIES

Generally, utilities have been involved with energy conservation programs only since the 1973 Arab oil embargo. During subsequent years, these programs and the motivations for offering them have undergone considerable change. Initially, programs were offered in an effort to maintain good relations with customers. As energy supplies became tighter and prices higher, some utilities viewed conservation programs as a customer service, assistance to help customers cope with higher energy prices. Some utilities developed programs in response to state and federal regulatory requirements. The responsive nature of these programs resulted in measurement and evaluation being neglected. Because these programs were not within the utility's mainstream activities and were often considered public relations (or public education) and/or regulatory response, utilities saw little need to subject these programs to rigorous cost-effectiveness considerations. Generally, these programs were based on little prior consumer research and/or cost-benefit analysis (Electric Power Research Institute 1984a).

Since these early beginnings, however, some utilities (especially electric utilities) have begun to offer demand-side (or customer-side-of-the-meter) conservation programs to reduce the overall costs of providing energy services (e. g., to minimize the cost of heat, light, and refrigeration). These efforts mark the beginning of a new era for utility conservation programs in which "conservation energy" resources are being incorporated within the utility's traditional power-supply planning framework.

Inclusion of demand-side management programs in a utility's business portfolio offers three major potential benefits. First, purchase of conservation and load-management resources is often less expensive than purchase of traditional energy supplies. Second, these demand-side programs may improve relations with both customers and regulators. Third, the much smaller size of energy conservation services (relative to traditional generation facilities) allows utilities more flexibility in terms of capital costs, financing requirements, environmental quality, timing and scale of activities, and regulatory approval. However, implementation of demand-side programs requires utilities to develop new and perhaps unfamiliar skills and to undertake programs with considerable uncertainty.

During the 1970s both gas and electric utilities conducted conservation programs. More recently, gas utilities have enjoyed adequate supplies of natural gas but have experienced reduced sales because of residential conservation, industrial substitution of other fuels for gas, and increased competition from electric utilities. As a result, in the 1980s most of the new conservation program development is being initiated by electric utilities. Review of the structures of the two industries illustrates the differences in their business objectives.

Besides the differences between gas and electric utilities, substantial variations exist among individual utilities. Factors that affect the potential value of customer conservation actions include reserve margins; gas supply or electric-generating plant fuel mix; load factors; peak-demand and energy growth rates; distribution of sales by customer class (residential, commercial, industrial); utility financial condition and capital market status; external energy-purchase requirements and availability; siting issues for proposed new power plants; and customer relations (Electric Power Research Institute 1984a). Each utility has to evaluate conservation programs in terms of its own system characteristics.

ELECTRIC UTILITY ASSESSMENT OF CONSERVATION

Whether a utility has enough, too much, or too little capacity significantly influences its marginal costs of energy. A study of 13 electric utilities with innovative conservation programs revealed enormous variation in the marginal costs for additional energy supplies (Electric Power Research Institute 1984a). Marginal costs are the yardstick against which the costs of conservation programs are evaluated for their potential benefit to the utility. In terms of capacity, marginal costs ranged from $370/kW for a combustion turbine (peaking plant) to $3,000/kW for a nearing-completion nuclear plant; the median cost was $1260/kW. Marginal energy costs ranged from $0.019/kwh to $0.12/kwh with a median of $0.0932. These figures show the substantial variation in the economic desirability of conservation and load-management programs to utilities.

A utility that requires new capacity with high capital costs is likely to pursue energy conservation opportunities aggressively because of the high cost and the long delays associated with completion of a major new plant. A utility with adequate capacity except at peak times may value load-shifting and control that affects energy uses at peak times. A utility with excess capacity may have no interest in conservation, may offer only small conservation programs either as a customer service or to gain experience for potential use in the future, or may start a marketing program to expand the use of electricity (Brown and Levett 1984).

Careful analysis of both the short- and long-term impacts of conservation programs is essential. Such analysis might indicate that such programs are economically attractive even for utilities that currently have excess capacity. That is, the short-term revenue loss during the present surplus period might be more than offset by the ability to defer construction of additional expensive generating plants.

GAS UTILITY ASSESSMENT OF CONSERVATION

Gas utilities have an entirely different outlook on energy efficiency. Most gas distribution utilities buy their supplies from others. Depending on the market for gas, they may have uniform costs or marginal costs for supplemental sources of gas as much as 50% higher than their average costs (Wirtshafter 1983). Generally, these gas commodity costs are passed along to ratepayers through adjustment clauses with oversight by state regulators. Unlike investment in new electric generating plants, gas utilities do not make major capital expenditures when demand expands.

Gas utilities are faced with two potential economic problems: gas sales have fallen during the past decade, and existing and potential new gas markets face heavy competition from electric utilities and efficiency improvements to electric appliances. The gas industry has a serious challenge to its continued profitability, and energy efficiency may not fit in. However, the Gas Research Institute devotes a substantial share of its budget to R&D projects intended to improve the energy efficiency of new gas HVAC equipment and appliances. Thus, GRI and the gas industry view improved equipment as an important part of their marketing efforts to maintain and expand their market shares.

Gas usage for residential appliances dropped substantially from 1975 to 1982 (American Gas Association 1984). Consumers use 29% less gas for furnaces, 18% less for water heaters, and 25% less for cooking ranges. Decreased sales were caused by shortages of gas during the mid-1970s, price increases, conservation actions, and replacement of inefficient equipment. The relative contributions of each are unclear, but the overall impact was to decrease the sales and profits of gas utilities.

As gas prices rose in the early 1980s, gas utilities lost growing amounts of gas sales as industrial users turned to other fuels. Since industrial sales account for almost half of U.S. gas utility revenues, this decline seriously eroded the economic position of many gas utilities. The response of the gas industry has been to establish new, competitive gas rates for industrial customers, to increase the penetration of efficient gas appliances in new residences, and to offer some conservation programs to keep customers from replacing gas equipment with electric products. The willingness of gas utilities to spend money for conservation programs, therefore, depends on an individual gas utility's competitive position vis-a-vis the local electric utility and the prospect for an expanded industrial market to buy the gas conserved by residential customers.

Combined gas and electric utilities face the same issues as those faced by separate gas and electric utilities. The difference is that the costs, benefits, and tradeoffs will be analyzed within one organizational framework rather than in separate organizations. Whether or not an optimum overall strategy is selected will depend on the combined utility's cost-accounting approach, management philosophy, and regulatory direction.

An example illustrates the difficulty gas utilities face. Baltimore Gas and Electric (BG&E, a combination utility) performed a detailed analysis of the economic attractiveness of conservation and solar actions by customers through both a regular Residential Conservation Service (RCS) program with enhanced financing services and a limited RCS program (Morgenstern and Dubinsky 1983). The enhanced RCS program involved a 10-year program of audits, subsidized financing (alternative subsidy levels were evaluated), and extensive technical assistance. Depending on program design options, BG&E estimated net-present-value costs to the utility of $1 million to $96 million on the electric side and $3 million to $44 million on the gas side (its purchase contract has no marginal gas costs, and BG&E assumed no increased industrial market for conserved gas) for a total *loss* of $4 million to $140 million.

Concerned with the size of these costs, the utility evaluated a limited program targeted at energy conservation measures that would most reduce BG&E's summer electric peak, using a simpler audit and rebates rather than loans to cut administrative costs. This limited program would still have a net cost to the utility of $8 million to $30 million, or up to $23 per residential customer. At this level of utility cost, the program might be supported as a contribution to the community and certainly to the participants, at only a modest cost to non-participating ratepayers. Two other utilities came to similar conclusions: Northeast Utilities (a combined utility), whose residential low-cost weatherization program is available to oil-heating homes as well, and the Palo Alto Utility (a municipal combined utility), where program costs are justified as a service to the community and homeowners.

REGULATION OF UTILITIES

The regulatory process in each state is a key factor in how utilities value conservation programs. Public utility or service commissions (PUCs or PSCs) and state energy offices play a significant role in setting the requirements imposed on utilities and the financial rewards authorized for utility operations. In some states, conservation programs have been imposed on utilities by PSCs. This can lead to poorly designed and executed programs when market research or utility planning is absent.

An alternative is for commissions to establish "rewards" for utility-initiated conservation programs. These may come in the form of an upward adjustment in the allowed rate of return. The utility's rate of return can be cut if the utility fails to meet conservation objectives set by the commission. For example, Southern California Gas Company was awarded a $5 million bonus in 1982 for exceeding conservation goals established by the California PUC in 1980. The utility had been forewarned in 1980 that it could expect a reward or penalty depending on its performance (Lucchi 1984). The California PUC penalized Pacific Gas and Electric $14 million in 1980 and 1981 for insufficient progress with contracts for cogeneration and small-power production projects (Flaherty 1984). By 1984, PG&E had contracted for approximately 3000 MW of small power production.

The regulatory process today is neither elegant nor straightforward. On the one hand, properly motivated utilities struggling to ensure the "right" outcomes from regulatory proceedings must devise their strategies carefully. Regulatory approval of utility-initiated demand-side programs often hinges on careful preparation of evidence justifying those programs. This preparation requires advance planning and early contact with regulators and their staff prior to formal proceedings (Davis et al. 1984). On the other hand, well-intentioned regulatory directives have been sabotaged in their implementation by utilities when utility corporate objectives were at odds with regulatory direction.

Several PSCs have given broader mandates to utilities to engage in energy conservation and related efficiency activities, such as cogeneration. The Pacific Northwest Power Planning Council (a quasiregulatory body) and states like Florida, New York, Nevada, and California gave clear directives to utilities to develop wide-ranging energy conservation plans for review and authorization by PSCs.

A new concept for utility roles in energy efficiency was suggested by Sant (1979) called the "Least-Cost Energy Strategy." He suggested that utilities stop thinking of themselves as selling only traditional energy commodities and start thinking of marketing "energy services." These services are the comfort and productive services in buildings and not therms of gas or kilowatt-hours of electricity. Under the least-cost approach, the utility would have the means and the profit motivation to invest in energy efficient technologies and services if they were less costly than supplying customer energy requirements with conventional resources.

This proposal provides a way for utility investment decisions to become virtually synonymous with societal cost-benefit determinations.

Recently, the least-cost approach received significant regulatory approval. The National Association of Regulatory Utility Commissioners adopted a resolution (No. 7) at their 1984 annual meeting recommending state and federal policies that base "determination of need for new [energy supply] facilities on the development of a least cost supply plan which evaluates and incorporates all cost-effective conservation, load management and alternate energy sources in the least cost supply plan."

A primary objection raised to utility energy service companies comes from product manufacturers, contractors, consulting engineers, and others who fear the potential competition with utilities. Utilities have access to customers and detailed information on energy use, which might be used to the disadvantage of competitors in the energy service industry. The common adjectives used are "anticompetitive" and "discriminatory."

A variety of state and federal laws inhibit or prohibit utility investment on the customer side of the meter. These regulatory and legal constraints are based on concerns that utilities might operate programs in an anticompetitive fashion (U.S. Congress, House Committee on Small Business 1984). However, reviews by the Federal Trade Commission of utility operation of their RCS programs show almost no such behavior. If these restrictions are removed, utilities might be more willing to establish end-use investment and sales programs (Department of Energy 1984a and b).

Although progress has been made, the regulatory tools currently in use appear insufficient to ensure that utilities will seek a "least-cost" energy resource plan. The potential for utilities to play an important role in achieving "least-cost" energy services clearly will require both appropriate regulatory signals to direct utility planning objectives and a related entrepreneurial spirit within utilities to offer energy efficiency services to their customers.

MATURATION OF UTILITY CONSERVATION PROGRAM PLANNING

Recent stability in fuel prices and the growing understanding of the potential effects of conservation programs led to an evolution in how utilities view energy conservation. Utility conservation programs are becoming energy management programs. Electric utilities with summer peaks advertise the benefits of efficient heat pumps for winter heating (to build load factors) or offer rebates for purchase of high efficiency air conditioners (to reduce peak loads). Bonneville Power Administration,

Pacific Gas and Electric, and other utilities now include electric conservation and load reduction in their long-range plans. Other utilities are adding free distribution or direct installation of conservation products to their residential conservation programs to guarantee savings. Some utilities, such as those in California, the Pacific Northwest, and the Tennessee Valley, had been operating aggressive and effective conservation programs since the mid-1970s.

These changes are evidence that utilities are beginning to approach energy efficiency in a strategic way. Utilities are developing the capabilities to describe how conservation programs affect customer demand, sales, and revenues (Fig. 10.1). Conservation program budgets also have expanded to the point where they are noticeable, and hence need to conform to established principles of planning and management.

A survey of 120 electric utilities conducted by the Investor Responsibility Research Center (1983) found that almost three-fourths of the utilities had established "formal" energy conservation and/or load management programs. Altogether, the 64 utilities that provided cost data spent $864 million on these programs in 1982.* Projections of spending levels on these programs indicated substantial increases; the 33 utilities that provided cost projections showed an average increase in their expected annual budgets of about 50%. These figures provide further evidence of the importance that utilities currently attach to their efforts to affect decisions on the customer side of the meter.

By the early 1980s, many utilities had seen their costs for new plants and fuel supplies increase substantially in response to world energy markets, inflation, tightening capital markets, and growing difficulty in obtaining approval for new plants. Those utilities with substantial service-area growth and without the capacity to serve this growth began to look for alternative ways to meet future energy demands. For many utilities, energy conservation and load management could contribute a partial solution. This marked the beginning of an era during which utilities viewed conservation as an electric system resource, and no doubt it put a new light on the way utility management viewed conservation programs.

Utility management and state energy regulators began asking that conservation programs be "evaluated" in economic terms so conservation programs could be compared with the marginal costs of new generating capacity, the relatively high costs of operating peaking generation plants,

---

*Utilities spend about $150 million per year on the RCS program alone (U.S. Department of Energy 1984a); California's investor-owned utilities spent $309 million on all their conservation programs in 1983 (California Public Utilities Commission 1984).

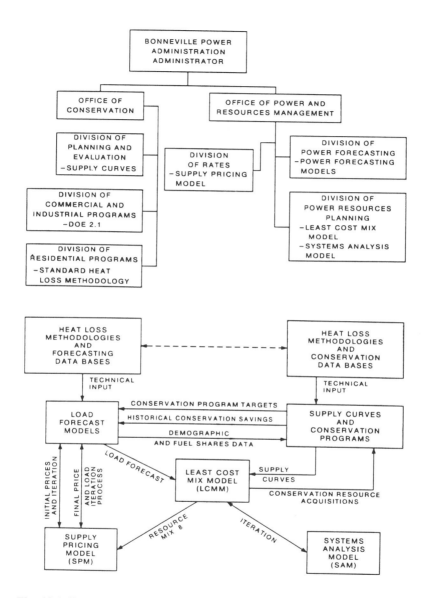

Fig. 10.1 Several divisions work on the planning, analysis, and implementation of BPA conservation programs (Tonn, Holub, and Hilliard 1986). The top portion of the figure shows the major BPA divisions involved with demand-side planning in BPA. The bottom portion shows the models used in this process and their interactions with each other.

the expense of purchasing and transmitting electricity from other utilities, and the rising costs of delinquent accounts of customers unable to pay fuel bills. Knowledge of these costs allows conservation programs to be considered as substitutes for traditional resource investments or as insurance to help utilities manage the risks inherent in long-range energy-demand forecasts and capital expansion plans (Sanghvi 1984). Lack of sufficient detailed data on *actual* program performance hinders adoption of new programs and expansion of existing ones (Morgenstern and Dubinsky 1983). Absence of such data (on actual energy savings, load reductions, and program operation costs) makes it difficult to adequately compare supply-side and demand-side programs with each other. In addition, this lack of data increases the already high levels of uncertainty (and perhaps discomfort) that traditional utility managers have with demand-side programs.

Utilities are now assessing how "demand-side mangement" alternatives fit with their "strategic planning" or resource planning, which in the past focused on supply and financial planning. By adding demand-side alternatives, the utilities formally recognize the potential to assess and influence energy use on the customer side of the meter to best meet the overall objectives of reliable and cost-effective energy services for their customers. Conservation, thus, has been placed within a planning system, where it has both the potential to compete and the responsibility to be evaluated against supply options (Naill and Sant 1984; Sawhill and Silverman 1983).

C. UTILITY PROGRAM EXPERIENCES

PLANNING DEMAND-SIDE MANAGEMENT PROGRAMS

The utility industry has six generic load-shape objectives to describe potential demand-side programs (Electric Power Research Institute 1984b). Only two, peak clipping and strategic conservation, are exclusively conservation-oriented (Fig. 10.2). The others have the potential to achieve efficient utilization of energy resources (in an engineering sense), but their implementation will not necessarily diminish consumption. For example, strategic load growth could involve switching customers from gas space heating to electric heat pumps in a summer-peaking utility, to improve the load factor. However, the same objective could be met by selling electric heaters for automobile engine start-ups on cold winter mornings (not an improvement in energy efficiency).

PEAK CLIPPING, or the reduction of the system peak loads, embodies one of the classic forms of load management. Peak clipping is generally considered as the reduction of peak load by using direct load control. Direct load control is most commonly practiced by direct utility control of customers' appliances. While many utilities consider this as a means to reduce peaking capacity or capacity purchases and consider control only during the most probable days of system peak, direct load control can be used to reduce operating cost and dependence on critical fuels by economic dispatch.

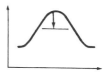

VALLEY FILLING is the second classic form of load management. Valley filling encompasses building off-peak loads. This may be particularly desirable where the long-run incremental cost is less than the average price of electricity. Adding properly priced off-peak load under those circumstances decreases the average price. Valley filling can be accomplished in several ways, one of the most popular of which is new thermal energy storage (water heating and/or space heating) that displaces loads served by fossil fuels.

LOAD SHIFTING is the last classic form of load management. This involves shifting load from on-peak to off-peak periods. Popular applications include use of storage water heating, storage space heating, coolness storage, and customer load shifts. In this case, the load shift from storage devices involves displacing what would have been conventional appliances served by electricity.

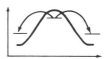

STRATEGIC CONSERVATION is the load shape change that results from utility-stimulated programs directed at end use consumption. Not normally considered load management, the change reflects a modification of the load shape involving a reduction in sales as well as a change in the pattern of use. In employing energy conservation, the utility planner must consider what conservation actions would occur naturally and then evaluate the cost-effectiveness of possible intended utility programs to accelerate or stimulate those actions. Examples include weatherization and appliance efficiency improvement.

STRATEGIC LOAD GROWTH is the load shape change that refers to a general increase in sales beyond the valley filling described previously. Load growth may involve increased market share of loads that are, or can be, served by competing fuels, as well as area development. In the future, load growth may include electrification. Electrification is the term currently being employed to describe the new emerging electric technologies surrounding electric vehicles, industrial process heating, and automation. These have a potential for increasing the electric energy intensity of the U.S. industrial sector. This rise in intensity may be motivated by reduction in the use of fossil fuels and raw materials resulting in improved overall productivity.

FLEXIBLE LOAD SHAPE is a concept related to reliability, a planning constraint. Once the anticipated load shape, including demand-side activities, is forecast over the corporate planning horizon, the power supply planner studies the final optimum supply-side options. Among the many criteria he uses is reliability. Load shape can be flexible — if customers are presented with options as to the variations in quality of service that they are willing to allow in exchange for various incentives. The programs involved can be variations of interruptible or curtailable load; concepts of pooled, integrated energy management systems; or individual customer load control devices offering service constraints.

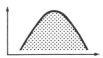

Fig. 10.2. Electric utilities can adopt several strategies to affect the size and shape of customer loads (Electric Power Research Institute 1984c).

Utilities' current approach to evaluation of demand-side program options is to apply the same production-cost models used for generation planning and scheduling. These models require data on the magnitude and timing of customer demand for electricity, the load shape. With load-shape information, production-cost models can then be integrated with financial-analysis models to assess the financial impacts of alternative supply and demand strategies (Caldwell 1984). Demand-side programs therefore must be able to quantify effects in terms of kWh, time of use, and season of the year.

Major issues facing demand-side planners include accuracy and reliability of data about specific demand-side options, the potential for double-counting of conservation impacts resulting from nonprogram factors, consumer decision-making processes, and the importance of risk beyond the control of the utility.

Although more than 1000 demand-side utility projects are underway in the U.S. (Electric Power Research Institute 1984b), the quality of data on the impacts of these programs is poor. The range of data problems includes:

- Limited load and submetered data on end uses of energy (especially in the commercial sector) and the impacts of conservation actions,
- Limited information on the current stock of equipment and available technologies,
- Uncertainty about customer acceptance of and response to conservation programs and technologies,
- Ignorance of the response to different implementation strategies, and few market potential studies to identify the range of effects that might be obtained (Davis et al. 1984; Electric Power Research Institute 1984a).

Utilities need to develop a substantially greater data base in three areas: energy end-use (therms, kwh, and kw demand), customer decision-making processes, and the impacts of program strategies and their component elements. End-use information requires knowledge of both existing technical efficiency and the market potential for new products and systems disaggregated by market segment and facility type (single-family vs multifamily; restaurant vs office building). The decision-making process must be understood if utility, manufacturer, and energy management services are to successfully appeal to the factors influencing energy-use decisions. Information on the effects of program strategies and techniques is critical to reaching higher percentages of the market potential for efficiency.

Fortunately, the increasing size and importance of these programs has led, during the past few years, to greater attention to data collection, program planning, and program evaluation. Many utilities, state regulators, and DOE have recently conducted useful evaluations of utility conservation programs. For example, some researchers have begun to distinguish between naturally occuring conservation and program-motivated actions for residential weatherization programs (Hirst 1984; Kushler 1984; O'Keefe 1983; Argonne National Laboratory 1985). However, more work is needed to better distinguish between conservation that would not have occurred without these programs and conservation that would have occurred anyhow. Moreover, little information exists to assess programs addressed to commercial and industrial customers.

A problem with demand-side analysis is that end-use metering is expensive, and many utilities avoid end-use metering (to the detriment of program certitude) through deductive analysis of building components, consideration of utility bills, and use of others' data. As an example of the costs involved, Sierra Pacific Power intends to spend $3.5 million to obtain metered data on energy end-use in 280 commercial buildings (Caldwell 1984). Some utilities use a combination of statistical methods, engineering analysis, and limited end-use load metering to infer load shapes at lower costs than large load-research projects would require.

To accurately forecast consumer response to new programs, utilities must obtain relevant information on consumer decision processes. They must also develop improved analytical methods to organize and use this information. Some attempts have been made to assess behavioral attributes of customers in terms of identifying market potential and marketing techniques. Notable among these is a model developed by Arthur D. Little and applied to Portland General Electric's zero-interest weatherization program (O'Keefe 1983). The model is designed to answer questions, such as who will participate, how much will they save, and what will the impact(s) be on the utility. Similar analyses were performed by ORNL on residential retrofit programs (Hirst 1984) and by the Southern California Edison Company to determine marketing techniques (Williams 1983).

Another approach to assessing expected program impacts is to conduct formal program experiments. Experimental evaluation of alternative program implementation strategies may be expensive, time-consuming, and fraught with concerns about equal treatment of parties offered or denied the service. However, carefully designed *small* experiments can provide valuable information on the effectiveness of program alternatives and variations (e.g., comparisons of different ways to market a residential appliance-rebate program, of different energy auditor training methods, and of rebates vs loans). Properly conducted experiments can provide reliable information that is unavailable anywhere else.

Variations (either within a single utility or among several utilities operating similar programs) can be used to assess relative program effectiveness. For example, BPA's incentive program is aimed at intensive energy retrofits from a one-time program contact. Pacific Gas and Electric offers commercial incentives to get customers started with specific measures now in the hope of motivating additional conservation measures over time. Southern California Edison operates a commercial incentive program tied to an energy audit to spur actions by those customers who have a utility audit. These three programs could be evaluated together to determine the effectiveness of different implementation strategies.

Recognizing both the importance and the costs of collecting relevant data, EPRI embarked on a three-year, $1.6 million demand-side management project (Electric Power Research Institute 1984c). The objectives of the project are to develop the methodological tools, data bases, sources of outside data, and planning techniques necessary to bring demand-side planning closer to the analytical sophistication used for power-production planning (Electric Power Research Institute 1984b). One of the major tasks of this project is to determine how data sources and program strategies can be transferred among utilities to save development time and the expenses of data collection and interpretation by individual utilities.

One of the most difficult tasks facing utility demand-side planners is sorting out the contributing factors to changes in energy use. In addition to such factors as price and consumer discount rates (discussed earlier), multiple utility conservation programs or activities complicate assessment of impact on consumer actions. A survey of 13 utilities (Electric Power Research Institute 1984a) found that these utilities offered 187 conservation programs, of which 125 were directed to residential customers; in many cases these strategies worked in combination to influence consumer actions. As a result, no one program was responsible for any change in energy use. The interrelation of programs should be kept in mind in evaluating programs because it may be appropriate to perform a cost-effectiveness analysis on a set of integrated programs rather than on a single program.

The no-losers test is sometimes applied to conservation and load-management programs. This test requires that the utility's cost to operate the program plus the lost revenue caused by reduced energy use and loads from program participants be less than the cost of energy supply. If this test is met, then the utility can lower rates to all its customers. However, many programs provide benefits to participants that are larger than the utility's lost revenue; in these cases society as a whole benefits from the program even though nonparticipants face higher rates. Considerable controversy surrounds application of the no-losers test both because it may eliminate programs that are worthy social investments and because no comparable test is considered for energy *supply* expansion programs.

INFORMATION PROGRAMS

The most common utility conservation program is provision of information to customers. Programs vary from general purpose media advertising to brochures to RCS home energy audits. Information and technical assistance programs are offered to commercial customers as well. Commercial-sector programs are more apt to focus on specific conservation technologies (e.g., lighting or energy management systems) or to be addressed to groups of like customers, such as restaurants or retail stores.

Information programs build a general awareness of how energy is used and what actions a consumer might take to reduce consumption. Utilities are often viewed as a credible source of information. Such programs, however, have little direct effect on the conservation actions customers take (Collins et al. 1985). A detailed review of innovative utility programs (Electric Power Research Institute 1984a) concluded that information programs were generally effective in building awareness of programs but did not produce measurable savings. The study observed a utility trend to combat this limitation by increasing promotional emphasis on the purchase of specific energy-conserving hardware (which presumably can be measured). Also, little evidence exists about the application of communication theory to energy conservation marketing strategies or to the design of specific, targeted, customer communications (Stern and Aronson 1984).

Some attempts have been made to design information services that strongly motivate consumer action. One example is the "report-card bill," which indicates energy use for the current month and for the same month a year ago. This information gives the customer positive feedback if he or she has taken conservation actions in the past year. This concept can be extended by forecasting what next year's bill will be if no further conservation occurs. The Palo Alto Utilities Department applies this concept in their multifamily and small commercial audit reports by estimating both the current year's and future years' energy savings, including a forecast of price increases for gas, electricity, and water. A detailed home energy report, designed by Fels and Kempton (1984), gives the customer a weather-normalized comparison of annual energy use and expenditures for the past year and a relative index of energy use compared with the state average for past years.

The effectiveness of post-audit programs in motivating conservation action is important. Evaluations of the RCS program (Chapter 9) conclude that program enhancements added by some utilities (such as low-interest loans, collaborative marketing with contractors, arranging for installation by qualified contractors, and establishing quality standards) have been responsible for greater energy savings than achieved with an information audit alone. These enhanced RCS services add to the cost of utility programs, however, and must be carefully evaluated in terms of cost-effectiveness.

INCENTIVE PROGRAMS

Four general forms of utility incentives are rates, loans, rebates, and free installations.

Rates

Utility rates can be structured to give energy users clear signals that conservation will be rewarded by utility bill savings. Designs that accomplish this include inverted rates (where prices are higher for greater levels of consumption), discount rates (where lower rates are offered to buildings meeting certain conservation requirements), time-of-use (TOU) rates (where rates are higher during the utility's peak hours or season, corresponding to higher generating costs), high demand charges (where charges are determined by the highest instantaneous customer demand for power, regardless of how long that demand lasts), and interruptible rates (where customers must reduce their use when so requested by the utility). Some of these rates affect energy efficiency, others affect only the timing of energy use, and some affect both consumption and demand (Table 10.1).

Duke Power Company offers a special conservation rate to occupants of new, well-insulated homes having efficient electric space heating systems (Electric Power Research Institute 1984a). The rate lowers electric charges 12 to 14%. Duke can offer this rate because, by meeting high energy efficiency standards, the homeowner effectively reduces the capital and operating costs of the utility.

Time-of-use or time-of-day rates are used for many large commercial and industrial customers because such customers contribute heavily to

Table 10.1. Electricity rate structures used to improve efficiency of electricity generation and use

Demand Rates	rates based on the maximum kilowatt usage of a customer, thus providing an incentive for customers to improve their load factor.
Time-of-use Rates	variable rates with higher per unit costs to the customer for use during a utility's peak periods and lower costs during off-peak periods.
Inverted Rates	rates where consumers pay more for each unit of electricity consumed in later tail blocks; the first block may or may not consist of a lifeline rate.
Seasonal Rates	rates that differentiate among the seasons in which the energy is consumed; with the season in which the utility reaches its peak having a higher rate than other seasons.
Variable Levels of Service	customers subscribe to a minimum electric service that satisfies their needs, e.g., interruptible rates.
Promotional Rates	rates designed to attract targeted groups of customers to a utility service area.
Off-peak Rates	rates priced to reflect lower off-peak costs, for specific end uses such as storage heating.
Conservation Rates	reduced rates based on a customer's dwelling meeting minimum energy efficiency standards.

Source: Electric Power Research Institute (1984b).

most utilities' demand and because metering equipment is affordable. TOU rates have been tested for residential and smaller commercial energy users. In 1975 and 1976 the Federal Energy Administration (now DOE), utilities, and PSCs carried out joint experiments with residential time-of-use rates (Faruqui et al. 1981). The conclusions drawn from twelve of these experiments are as follows (Faruqui and Malko 1983):

- TOU rates generally reduce peak-period electricity consumption. A Wisconsin experiment found that reductions during a utility's summer peak ranged from 15 to 32% and during the winter peak ranged from 4 to 16%. Large reductions occurred where the rate incentives were higher (e.g., 4:1 or 8:1 ratios of peak to off-peak rates) and the periods of control shorter (e.g., 6 hours per day vs 9 or 12 hours per day) (Caves and Christensen 1983).

- Responsiveness to peak and off-peak electricity prices ranges considerably, varying according to total electricity usage, appliance inventories, climate, and price differentials.

- No new peaks are created immediately before or after the peak hours.

- Overall electricity consumption generally declines.

Another form of rate incentive is the Demand Subscription Service tested by Southern California Edison Company. In such a program, participating households agree to keep electrical demand below a subscription level in return for lower monthly rates. The utility activates a control device when a system peak is anticipated. This device limits participants' electric demand to subscription levels by turning off the entire house when the subscription level is exceeded. Peak-period demands were reduced by an average of 1.3 kW per customer on an average day and by 2.1 to 3.0 kW per customer on the system-peak day.

Loans and Rebates

Many utilities offer low- or no-interest loans to assist customers in purchasing energy efficient equipment or to weatherize their buildings. The objective is to reduce the first-cost barrier to investing in conservation and sometimes to offer a loan repayment plan in which energy bill savings exceed monthly loan payments. Experience shows that loan programs have high administrative costs and debt-service expenses for utilities. For these reasons, some utilities have phased out their loan programs or offer customers the choice of a loan or a rebate. (Some consumers prefer loans while others prefer rebates.)

Utility loan and rebate programs have proliferated. A survey conducted in 1983 showed that financial incentive programs are offered by utilities serving 60% of all U.S. households (Dickey et al. 1984). Residential heat pumps and air conditioners are the most common products for which rebates are offered, and these are offered primarily in southern states, such as Florida and Texas, with high summer peak demands (Dickey et al. 1984; Geller 1983). Other rebated appliances include solar water- and space-heating systems, heat-pump water heaters, efficient gas or electric water heaters, refrigerators, freezers, and gas ranges. The national survey indicated that peak-load reduction was the primary utility objective. The results of one such effort to reduce the summertime peak load are shown in Fig. 10.3. The second most frequent objective was "miscellaneous," offered primarily by gas utilities; these incentives were probably aimed at building gas-utility market share.

The Tennessee Valley Authority has offered zero- and low-interest loans in weatherization and heat-pump programs since their inception in 1977. As of October 1984, about 450,000 weatherization and 57,000 heat-pump loans had been made (Tennessee Valley Authority 1984). These

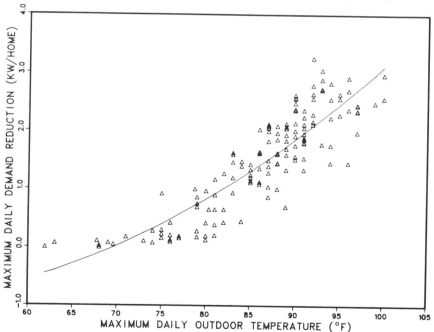

Fig. 10.3. On-peak air conditioning demand reduction as a function of outdoor temperature, Arkansas Power & Light (Kuliasha et al. 1985).

programs are considered successful because they reach a substantial fraction of the potential market, have reduced peak winter demand by more than 600 MW, and are cost-effective to both the TVA power system and consumers (Tennessee Valley Authority 1984).

Many other utilities also offer residential weatherization loans, largely in conjunction with RCS audit programs. Northern States Power Company in Minnesota found that those who took a loan saved an additional 15 MBtu (an additional 9% savings) over audit-only customers (Hirst 1984). A Bonneville Power Administration weatherization program had a similar finding (Hirst 1984). There, program-induced energy savings were zero for the audit-only households and 3500 kWh per year (a 13% net savings) for those receiving both audits and loans. Puget Sound Power and Light found net savings of 1000 kWh and 3800 kWh for audit-only and audit-plus-loan participants, respectively (Electric Power Research Institute 1984a). These studies show that addition of a loan to a home energy audit program increases program energy savings by a factor of about three.

Loan programs can be expensive, however. Pacific Gas and Electric's zero-interest loan program cost $95 million for 500,00 dwelling units. This cost, equivalent to $190 per unit, covers only administration and debt service (California Public Utilities Commission 1984).

Loan offers can also substantially increase program participation rates. The Snohomish, Washington, Public Utility District found that by offering the financial incentive paid by BPA, the utility had a 23% response to RCS audit announcements in two years (Glazer 1984).

Both Florida Power and Light and Puget Sound Power and Light have found they can offer customers a substantial (about 70%) cash rebate at the same cost as a zero-interest loan (Electric Power Research Institute 1984a). Such a direct rebate, however, has a slightly different rationale for encouraging conservation actions. It may be offered where the consumer might not take out a loan or where the consumer makes impulse buying decisions and is unlikely to apply for a loan. Finally, rebates can be less costly for utilities than loans.

Qualification levels and rebate payments vary from program to program. In most cases, fixed rebates are given. In a few cases, payments are made on a sliding scale according to the efficiency of the product. The sliding scale approach is preferred by equipment manufacturers because it provides an incentive for the purchase of highly efficient units. Utilities, however, prefer the administrative simplicity of fixed rebates (Geller 1983; Savitz et al. 1984).

Florida Power and Light (FP&L) designed each of its rebate programs to achieve a standard 5:1 benefit/cost ratio for the utility. Most of their programs meet this goal. For example, an incentive program for customer purchase of efficient central cooling and heating systems costs FP&L an average of $440 per qualifying installation. The rebate varies according to the energy efficiency ratio (EER) and the size of the equipment. For this program, the utility estimates it is paying about $500 per kW of summer peak demand reduction and $0.17 per kWh saved in the first year, thereby meeting the strict cost-effectiveness criterion (Elliott 1985).

Texas Power and Light pioneered the appliance-rebate concept as part of a company-wide goal to reduce peak demand by 1000 MW. The company offers rebates to those purchasing efficient air conditioners, heat pumps, and water heaters at a cost of about $150 per kW of load reduction. The program of rebates to builders and homeowners includes incentive payments to dealers as well.

Northern States Power (NSP) did extensive analysis of their appliance-rebate pilot program in 1982 and 1983 (Gunn 1983). About 26,000 rebates were paid for residential appliance purchases. Overall, NSP concluded that the cumulative impacts of the program through the year 2000 would produce a net benefit for efficient air conditioners, and a small penalty for efficient water heaters, freezers, refrigerators, and refrigerator-freezers. A net benefit would accrue to ratepayers for all appliances if the program were delayed several years. This conclusion corresponds to the utility's expectation that demand reductions have little value in the short run, but reduce the need to invest in new plant construction in the 1990s.

The NSP study produced some other interesting results. Almost 40% of appliance purchasers eligible for rebates did not bother to apply. Apparently the rebate was not that important, or people did not know about it. The rebate seems to make an impact at the last stage of the consumer's decision, when the final issues are availability and price; the rebate, of course, affects the latter. NSP concluded that a key element of success was getting appliance dealers to stock and promote efficient appliances (Gunn 1983).

General reviews of rebate programs have reached conclusions similar to those based on the specific programs described here (Geller 1983). However, even widely available utility incentive programs aimed at consumers have produced only a small direct effect on the average efficiency of new products sold in the U.S. (Dickey et al. 1984).

The success of rebate programs depends on close cooperation with "trade allies," the appliance dealers, manufacturers, and contractors who influence consumer purchase decisions. This cooperation may take the form of incentive payments to dealers and contractors, consultations to determine rebate levels, verification mechanisms that minimize paperwork for dealers (e.g., lists of eligible appliances and certification stickers), and planning to ensure that adequate supplies of qualifying products will be in stock (Electric Power Research Institute 1984a).

Commercial Sector Incentives

Some utilities have initiated commercial-sector rebate programs (Savitz et al. 1984). Products include energy management systems, lighting, air conditioning, heating, and some process equipment, such as motors, pumps, and refrigerating and cooking equipment. Major efforts have been launched by Pacific Gas and Electric (PG&E), which spent $32 million on rebates in 1983 (Calhoun 1984); Puget Sound Power and Light (Lehenbauer 1983); and Southern California Edison.

PG&E estimated that their program saved 100 MW of demand in 1983 alone and that they are buying electricity savings at 0.5 cents per life-cycle kWh (Calhoun 1984). However, PG&E does not account for efficiency investments that would have occurred in the absence of their financial incentives.

Puget Power developed its incentive program in conjunction with its commercial and industrial audit program. Incentives are offered selectively to those customers interested in conservation but needing some assistance to take action. Puget Power estimates their savings cost 0.5 to 1.0 cents/kWh on a life-cycle basis (Lehenbauer 1983). Although commercial sector incentive programs appear to be highly cost-effective, neither PG&E nor Puget Power includes overhead costs to administer the program nor the customer's share of the conservation investments.

Very little research has been conducted by utilities to compare different incentive schemes or rebate levels (Berry et al. 1983). Utilities generally have not tested different types of strategies by varying the amount of the incentives or by offering different incentives to separate market segments. Therefore, little is known about optimal levels and types of incentives.

Some innovation in commercial sector incentive programs is occurring. The Bonneville Power Administration operates a pilot program to buy "conservation savings" from five parties (ranging from individual customers to building owners, shared savings companies, and possibly another utility) at an estimated 2.0 cents per life-cycle kWh (Schick 1984). San Diego Gas and Electric is experimenting with a novel rebate program to encourage thermal energy storage for space cooling. They offer a variable incentive to customers willing to use cooling storage, such as chilled water, ice, or eutectic salts, to reduce summer peak demand. The incentive is in the form of a guaranteed three-year payback on the customer's additional investment in cooling equipment. This calculation factors in climate zone, rate schedule, building occupancy, and the type of storage system selected, all of which make the incentive much more performance-based than most incentives. The utility estimates that up to 85% of peak cooling electrical demand can be shifted off-peak with thermal energy storage.

Free Installations

After comparing the costs of marketing and supporting information, loan, and rebate programs to the resulting energy savings, some utilities have found it more cost-effective to offer free installations of some conservation devices. In the late 1970s, several utilities (such as Portland General Electric and Southern California Edison) offered free water-heater insulation plus installation to customers with electric water heaters. It was cheaper to go door-to-door installing these for free than to sell the insulation.

A comprehensive free installation program was run in Santa Monica, California, under contract to the local gas and electric companies (Chap. 9). For less than the cost of a utility-supplied RCS audit, the Santa Monica Energy Fitness program conducts a streamlined audit and installs three low-cost conservation measures (Egel 1984).

PG&E through the California/Nevada Community Action Association experimented with different ways of increasing participation of low-income households in weatherization programs (Moulton 1984). Offering no-interest loans was ineffective. Direct weatherization at no cost to the owner or resident turned out to be very effective in reaching low-income households, and had approximately the same cost per household for PG&E ($600 to $750) as no-interest loans because of the interest, outreach, and overhead costs for the loan program.

HYBRID ENERGY SERVICES

The Palo Alto Utilities Department conducts a competitive bidding process to allow local contractors to bid a material and installation price on insulation applicable to all homes participating in their program. MASS SAVE in Massachusetts offers a similar program, in which the RCS contractor works with local communities to arrange volume insulation contracts (McDonough and Parisi 1984). Northeast Utilities operates a very successful Wrap-Up/Seal-Up program, which contracts with a private company to install low-cost weatherization devices at nominal costs for Connecticut residents.

Where quality control has been a problem in installation of weatherization products, utilities have developed several services. These include standards required for contractor participation in loan or rebate programs, inspection of work, workshops to teach proper installation techniques, and support of specially trained crews working for local government or nonprofit weatherization programs.

The Wisconsin Gas Company established a pilot Shared Savings Program to replace existing residential gas furnaces. The company identifies customers with high enough heating loads to warrant accelerated furnace replacement and replaces the furnace with a high-efficiency model at no cost to the homeowner. The company owns the new furnace and maintains it for ten years. Gas savings are shared with the homeowner; the pilot program was operated on a breakeven basis. For eligible homeowners, gas savings were 35% of heating use ($400 per year). Although customers liked the program, a public relations problem occurs among customers with low gas usage who were turned away from the program. Also, the utility had many "free riders" that had not been anticipated, marketing costs were high, and heating contractors disliked the utility's foray into "their market" even though these contractors provided equipment and installation (Schrader 1984).

General Public Utilities (GPU) guarantees energy savings through its RECAP program. The utility contracts with private energy service companies to provide complete conservation services to selected electrically-heated households at no cost to the customer. The utility pays the company 5 to 8 cents per kilowatt-hour saved over several years. This program began in five areas in 1983. GPU estimates that, with 15,000 homes serviced in 10 years, net savings to the utility should amount to $85 million after expenses of $19 million (Esteves 1983; Brown and Reeves 1985).

Some utilities have assessed their experiences and the potential profit to be made from marketing energy services and have concluded that it is advantageous to establish unregulated, subsidiary, energy service companies to market energy efficiency directly to consumers. Energy service subsidiaries have been established by Northern States Power, Portland General Electric, and the New England Electric System.

D. SUMMARY

This discussion of utility conservation programs was, in many respects, quite similar to the prior discussion of government and local community programs. Not surprisingly, the findings are also similar.

Utility conservation programs have increased tremendously in scope, breadth, intensity, cost, and sophistication since the 1973 oil embargo. Early programs were primarily "soft" information and education efforts designed to increase consumer awareness of energy issues and their energy-efficiency options. Later programs supplemented these "advisory" efforts with much stronger site-specific information programs (on-site home and commercial-building energy audits) and financial incentive programs. These later programs focused more sharply on motivating consumers to adopt specific conservation and load-management technologies and to change specific energy management practices.

As utilities gained experience with these customer conservation programs, their view of the programs often changed. From initial perspectives varying from skepticism and cynicism to genuine concerns for customer service, a growing number of electric utilities now view demand-side management programs as legitimate business activities. Generally, gas utilities are much less active in demand-side programs because gas supplies are currently abundant and prices are stable. Electric utilities now judge demand-side programs by the same criteria used to assess traditional power supply programs: on the basis of their effects on revenue requirements, cost per kWh and per kW saved, and capital requirements. Increasingly, utilities are integrating their demand-side and supply-side planning activities. Most observers consider this a positive and healthy trend within the utility industry.

Unfortunately, little is known about the actual performance of these utilities' programs (as noted earlier for government and local community programs). That is, considerable uncertainty remains about the costs of these programs, customer participation rates, and the ultimate net effects of these programs on customer energy consumption and demand. For example, a utility considering implementation of a program to subsidize installation of high-efficiency air-conditioning equipment in commercial buildings has little information available to it on how best to design the program (marketing, level and type of incentive, types of equipment to include in the program), on the likely fraction of eligible buildings that will participate in the program, and on the likely effects of participation on annual electricity use and peak summer loads.

Clearly, more and better program evaluation is needed. Improved methods of disseminating conservation program experience among utilities, government agencies, and others involved in delivering conservation services are also needed. Additional research is needed to develop methods that allow reliable transfer of experience in one location to guide decisions in other locations. Finally, improved analytical methods are needed to better integrate supply- and demand-side planning tools. In particular, energy-demand forecasting models need to be expanded to include additional factors beyond the traditional economic variables of fuel prices and incomes. Factors like consumer attitudes and effectiveness of information programs need to be included.

That comparable uncertainties occur on the energy supply side must also be recognized. That is, a utility planning a new generating plant faces tremendous uncertainties concerning the future need for that power, regulatory delays, construction costs and schedules, and availability of sufficient financing at reasonable interest rates. Together, these factors make it very difficult to estimate the likely total cost of new facilities. However, the lack of such data does not delay planning for new construction, so the lack of complete data on demand-side options should not be used as an excuse to delay or avoid demand-side programs, either.

We close this chapter on a positive note. Utilities (especially electric utilities) are becoming increasingly committed to demand-side management programs. Utility budgets for conservation and load-management programs are about a billion dollars per year (comparable to the combined conservation expenditures of all local, state, and federal energy agencies). This is equivalent to the cost of about one large power plant. Integration of demand-side programs within the utility system planning framework is expected to encourage greater and more frequent consideration of least-cost energy strategies. And increased program evaluation (to reduce uncertainty) and greater support from state and federal regulatory commissions should result in increasingly effective utility conservation programs.

11

INDOOR AIR QUALITY

A. INTRODUCTION

The preceding chapters covered many of the important issues related to energy efficiency in buildings. However, time and space limitations as well as the interests and knowledge of the four authors prevented us from covering some other issues that are also important. Some of these unexplored topics include thermal storage, load management and related communication technologies, active solar space conditioning systems, the role of energy-demand models in conservation program planning, and indoor air quality.

The present chapter discusses indoor air quality (IAQ), a topic that has important public-health implications and that some consider the Achilles' heel of energy efficiency in buildings. Several recent studies found that indoor air-pollutant concentrations can approach or exceed ambient (outdoor) air limits (National Academy of Sciences 1981). This is important because most Americans spend 80 to 90% of their time indoors. Tightening buildings to reduce air infiltration and energy use can concentrate indoor air pollutants. Thus, a pre-existing problem with indoor air quality can be exacerbated by energy conservation efforts.

Table 11.1 summarizes the state of knowledge regarding indoor air pollution. After years of neglect by researchers and health officials, significant progress has been made in the areas of pollutant identification, pollutant behavior, and pollutant control. At the same time, our understanding of exposure levels, interactions, health effects, control options, and policy alternatives is far from complete. The federal government spends only a few million dollars per year on IAQ research and protection (compared to hundreds of millions of dollars per year on outdoor air.)

Table 11.1. State of current knowledge regarding indoor air pollution

Emission Sources
 Chamber studies done for several sources
 Few measurements under dynamic conditions
 Few studies of emission rates during normal use
 Lack of information about distrubution of sources within population
Dilution
 Understanding of basic components affecting air-exchange rates
 Measurement techniques available
 Site-specific models developed, but more-general application problematic
 Only limited information available on distribution of air-exchange
  rates in existing buildings
 Mixing inside buildings without mechanical ventilation systems not well understood
Indoor concentrations
 Survey-type data collected for some pollutants
 Applicability of survey data to entire building stock unknown
 Dilution and mechanical filtration typically assumed to be first-order
  determinants of concentrations
 Chemical and physical interactions, as well as removal rates, not well defined
 Little known about variations in both removal and penetration rates

Human activity patterns
 General features of population activity pattern known
 Insufficient information about variations in activity patterns with
  age, sex, socioeconomic class, employment status, location, and season
Exposures
 Relatively few studies of personal exposures to air pollution
 Limited data indicate poor correlation between outdoor concentrations
  and personal exposures
 Lack of suitable instrumentation limits application of personal
  exposure studies
 Distribution of exposures across the population and effects of energy
  conservation on indoor exposures not known
Health effects
 Irritant, toxic, mutagenic, and carcinogenic effects noted for many
  indoor contaminants
 Additional information needed on central-nervous-system effects
 Epidemiologic evidence of adverse health effects available for some
  pollutants
 Data on dose-response relation accumulated for a few pollutants
 Numbers, characteristics, and distribution of chemically sensitive
  individuals not known
 Information lacking on health effects of long-term, integrated
  exposures compared to short-term, peak exposures

Source: Spengler and Sexton (1983).

B. POLLUTANTS AND POTENTIAL HEALTH EFFECTS

Table 11.2 lists the major air pollutants that can be found in buildings, along with their sources, health impacts, and control techniques. The variety of potential pollutants, sources, and health impacts is large. One of the greatest areas of uncertainty concerns the health effects from low-level, long-term exposures to these pollutants. Nonetheless, studies show that air pollutants can pose a significant health hazard at the concentrations found in some buildings.

Radon, a colorless and odorless gas, is perhaps the air pollutant of greatest concern in residences. Indoor radon concentrations typically range from 0.1 to 3.0 pCi/L (pico-Curie of radiation activity per liter of air), although concentrations exceeding 50 pCi/L have been found in homes built on granite bedrock (Sachs et al. 1982; Weiffenbach 1982). It is estimated that a lifetime exposure to 1 pCi/L will increase the risk of lung cancer by 0.2% (National Council for Radiation Protection 1984), causing 1,000 to 20,000 cases of lung cancer in the U.S. per year (Nero 1983).

Formaldehyde is a known irritant emitted by ureaformaldehyde foam insulation, pressed-wood products, and other synthetic or manufactured materials. It is of particular concern in mobile homes, where pressed-wood products are widely used (National Academy of Sciences 1981). Although epidemiological data are inadequate to determine if formaldehyde is a human carcinogen, it is likely to pose this risk given the results of animal experiments (Ulsamer et al. 1984). ASHRAE adopted a recommended formaldehyde exposure limit of 0.1 ppm (American Society of Heating, Refrigerating, and Air Conditioning Engineers 1981; Janssen 1984). The U.S. Dept. of Housing and Urban Development set standards in 1984 to limit formaldehype in new mobile homes to 0.4 ppm (Federal Register 1984). A considerable debate continues over the level at which public exposure should be permitted for health protection.

A wide variety of organic compounds can also be present at irritating and/or unsafe levels in commercial buildings, and the number of cases in which workers in office buildings are reporting such symptoms as headaches, nausea, and eye or respiratory irritation has been increasing (Repace 1982). The source of a "sick building syndrome" problem can be very difficult to isolate (Stolwijk 1984). Furthermore, knowledge regarding the health risks posed by the diverse group of organic chemicals that office workers are now exposed to (Repace 1982) and regarding the quantities and health impacts of airborne organic compounds in homes is limited.

Unvented kerosene and gas heaters and stoves, tobacco smoke, and vehicles in attached garages are the primary sources of combustion products in the indoor environment. Evidence is mounting that the

Table 11.2. The health impacts, sources and control techniques for various indoor air pollutants

Pollutant	Health impact(s)	Sources	Controls
Asbestos fibers	Asbestiosis (lung cancer) Asbestos insulation	Asbestos ceiling tiles and fire-proof matts	If intact, do not disturb Professional removal Professional encapsulation (often hot water/steam pipe)
Bacteria and viruses	Colds, influenza, and viruses	Human sneezing, coughing, and breathing Areas of high moisture	Local ventilation in kitchen and bathroom Keep air conditioning filters clean
Carbon monoxide	Headaches Lung and heart ailments	Combustion (kerosene heaters, unvented gas stoves, tobacco smoke, wood smoke and motor vehicles)	Cease use of kerosene heater Local venting near site of combustion Limit smoking More-efficient combustion through tune-ups and proper operation
Formaldehyde	Eye and respiratory irritation Suspected carcinogen	Urea-formaldehyde foam insulation, pressed and manufactured wood products, carpeting, and drapes	Limit purchase of products containing formaldehyde Use sealants to encapsulate formaldehyde in wood products Additional ventilation Ammonia fumigation Remove urea-formaldehyde foam insulation from walls

Table 11.2. (continued)

Pollutant	Health impact(s)	Sources	Controls
Nitrogen dioxide	Can cause lung damage	Combustion (see above)	Same as for carbon monoxide
Organic chemicals (volatile)	Irritation some are toxics, mutagens, and/or carcinogens	Aerosols, cleaners, plastics, paints, and varnishes	Discontinue use of products identified as sources. Use in well-ventilated areas (local ventilation)
Particulates	Lung damage. Often vehicle for other toxic substances to enter lungs	Combustion (tobacco smoke is the primary source)	Same as for carbon monoxide; Also, use of air filtering devices
Radon	Lung cancer. Stomach cancer	Soil. Well water. Brick, concrete, and stone	Block or seal paths of soil gas entry in basement (or slab foundation). Vent basements and crawl spaces. Reroute basement water drain. Air circulation. Subslab ventilation. Charcoal adsorption filtering of water
Tobacco smoke (passive exposure)	Eye and respiratory irritation. Long-term effects may be more harmful	Cigarettes and other tobacco products	Limit smoking to areas with local ventilation. Use of air-filtering devices

Source: American Council for an Energy-Efficient Economy (1985).

respirable particulates and other pollutants found in tobacco smoke pose a significant health threat to nonsmokers (Repace 1982). The adverse health effects from the combustion products released by other devices (e.g. gas ranges) are much less certain.

Asbestos-containing products, such as ceiling and floor tiles and insulation, are widely used in commercial and institutional buildings. If asbestos fibers become airborne and are inhaled, lung and other types of cancers can result. Measurements in some buildings have shown airborne asbestos fiber concentrations greatly in excess of the levels generally considered safe (Sawyer and Spooner 1978). However, the overall public health risk is uncertain because asbestos is supposed to be immobilized in binding materials. A health hazard is present when the material is friable or it is disrupted and fibers are released.

C. DETECTION AND MONITORING

Instruments for identifying potential IAQ problems are available and are being improved (Lawrence Berkeley Laboratory 1984). Pollutant measurements can be made on an instantaneous basis (grab sampling) or continuously. Personal monitors for measuring an individual's exposure to radon, formaldehyde, carbon monoxide, sulfur dioxide, nitrogen oxides, and additional organic compounds were developed during the past 20 years (Treitman and Spengler 1984), and a number of simple and inexpensive "passive samplers" that do not require a pump have been developed. Such instruments make widespread pollutant monitoring feasible.

Many field studies of IAQ pollutants have been completed, particularly surveys of indoor radon and formaldehyde levels. Besides providing data on actual pollutant levels, some of these studies evaluate how various factors relate to the levels found. For example, radon measurements in 99 homes showed no correlation between radon levels and air exchange rates (Nero et al. 1983). A survey in the Pacific Northwest found that certain housing characteristics, such as the presence of a basement, are associated with higher indoor radon levels (Thor 1984). Also, the relationship between the ages and formaldehyde levels of mobile homes showed that formaldehyde concentrations decline over time (Housing and Urban Development 1981).

The dispersion and fate of potentially harmful substances in buildings is highly complex. Integrated air quality and infiltration models for buildings are still in an early stage of development (Macriss 1983). Further work on pollutant dispersion, interactions between emissions, infiltration and ventilation systems, and IAQ modeling is required.

Monitoring of indoor air pollutants is beginning to occur outside of the research community. As of 1982, 32 states had programs or persons responsible for evaluating exposures to indoor air pollutants in nonoccupational settings (Bernstein et al. 1984). So far, formaldehyde is the most commonly covered pollutant in these programs. In addition, private firms are beginning to offer IAQ tests.

D. IAQ AND ENERGY CONSERVATION

Historically, indoor air pollutants were diluted through air infiltration. Homes were leaky and high ventilation rates were maintained in commercial buildings. Recent surveys of hundreds of homes showed typical infiltration rates of 0.5 to 1.0 air changes per hour (ACH) (Lawrence Berkeley Laboratory 1984). In new energy-efficient housing, rates can be reduced to as low as 0.1 ACH through the use of vapor barriers, caulks, and sealants. This tightening of the building envelope can lead to higher levels of indoor pollutants. Studies of radon and formaldehyde levels in homes in Canada (Fig. 11.1) showed that pollutant concentrations can exceed generally acceptable levels in tight homes (Dumont 1984).

The extent to which indoor pollutants are a problem depends more upon the strength of the source than upon the ventilation level. For example, studies of radon in buildings (Nero et al. 1983) and of formaldehyde emissions from building materials (Meyer and Hermanns 1984) have shown that the variability in source strength is primarily responsible for the order-of-magnitude differences in indoor concentrations from one house to another.

Infiltration rates vary over a wide range as weather conditions change, and indoor pollutant levels follow the infiltration rates. So, beside wasting energy, maintaining a high average ventilation rate does not always ensure good IAQ. Many energy efficient homes do not have air quality problems, and many leaky homes have polluted air (Nisson and Dutt 1985).

Techniques for reducing ventilation rates (e.g., variable-air-volume systems) are being widely adopted in commercial buildings (Int-Hout 1984). Also, some building owners and operators are apparently closing off outside air vents (in violation of building codes) to lower energy consumption (Repace 1982). These measures can result in the building-related health problems discussed above. However, in some cases complaints may be caused by the thermal conditions (e.g., higher thermostat settings) rather than by inadequate ventilation (Int-Hout 1984).

E. CONTROL STRATEGIES

Indoor pollution can be controlled by a number of approaches, including source removal or modification, air cleaning, local ventilation, and mechanical ventilation with heat recovery. The number of control techniques and knowledge regarding their applicability and effectiveness is growing; but in general, work on control strategies is far from complete (Lawrence Berkeley Laboratory 1984; Anachem, Inc., and Sandia National Laboratories 1982).

In the area of source removal, manufacturers have reduced the formaldehyde emissions rates from pressed-wood products by a factor of ten or more in the past 15 years (Meyer 1984). In addition, certain paints, varnishes, etc. are effective in significantly reducing formaldehyde

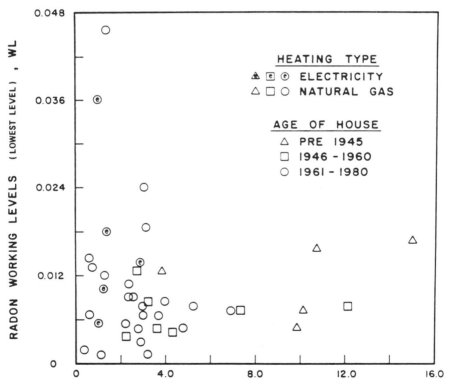

Fig. 11.1. Indoor radon levels as a function of infiltration rates in tight homes in Canada (Dumont 1984).

emissions, and fumigation techniques are available for short-term reductions (Lawrence Berkeley Laboratory 1984).

Techniques for reducing radon and radon-progeny levels in homes are also being developed (Sachs and Hernandez 1984; Scott and Findlay 1983; Nitschke et al. 1984). Sealing cracks in the basement walls and floor can prevent radon entry into the home; ventilation of crawl spaces or the soil beneath the basement can also be effective. However, further experimentation with and monitoring of radon-control techniques are needed.

The avoidance of heaters that vent their combustion gases indoors is an obvious way to reduce potential health hazards. Also, advanced gas stove burners that greatly reduce nitrogen oxide emissions are under development (Shukla and Hurley 1983). Another concern with fuel-fired appliances in tight homes is the use of indoor combustion air. It is now recognized that combustion air for major fuel-fired products should be drawn from outdoors to avoid such problems as backdrafts and oxygen depletion in residences (Nisson and Dutt 1985).

A variety of approaches to air cleaning are available, including filtration, electrostatic precipitation, ionization, and adsorption. These techniques are designed to remove particles from the air but not contaminants such as CO, NO_x, and formaldehyde. Some of the air-cleaning processes appear to be effective (Sextro et al. 1984), but further testing and product development are needed (Lawrence Berkeley Laboratory 1984).

Local ventilation can be very effective for providing fresh air and/or removing pollutants in particular rooms or areas. For example, use of a range hood with a gas stove can reduce the amount of combustion products entering occupied space by 40 to 90%, depending on the flow rate and details of the installation (Macriss and Elkens 1977; Traynor et al. 1982).

Mechanical ventilation in a tight house involves controlling the supply of fresh air and the exhaust of stale air, often with heat recovery between the two air streams. A number of companies in the U.S. are now producing air-to-air heat exchangers (Fig. 11.2) for performing this function (National Center for Appropriate Technology 1984). These devices can be wall or window mounted for ventilating part of a house or can be connected to the duct system for ventilating an entire home.

Recent evaluations have found some problems with the performance of air-to-air heat exchangers, however. Field studies in Canada showed that ventilation and heat recovery efficiencies were much lower than manufacturers' claims (Ontario Research Foundation 1984). Also, many homeowners who installed them do not operate them because of undesirable drafts and noise problems (Riley 1984). Even if air-to-air heat exchangers are working properly, they can be difficult to justify on the basis of the economic value of the recovered heat (Fisk and Turiel 1982; Macriss 1983).

Mechanical exhaust ventilation is an alternative to air-to-air heat exchangers, commonplace in new homes in France and Sweden. Exhaust ventilators can flush air out of homes with or without heat recovery. In about 70% of the new houses in Sweden, a water- or air-heating heat pump is used to recover heat from the exhaust air stream (Nisson 1984).

As of 1985, very few mechanical ventilation systems were in use in U.S residences, although one company was developing a forced-ventilation system with a heat pump for heat recovery. While it is clear that a house with mechanical ventilation can provide better IAQ than would be available in a conventional house (Nisson and Dutt 1985), much needs to be learned before mechanical ventilation becomes an ordinary feature of our homes.

Ventilation techniques in commercial buildings, particularly for reduced air-flow rates, also carry uncertainties (Janssen 1984). Unresolved issues include system efficiencies, air mixing, and bypassing related to the location of supply and return air ducts.

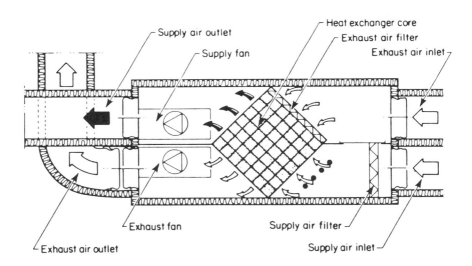

Fig. 11.2. Mechanical ventilation system wtih air-to-air heat exchanger for residential applications (Lawrence Berkeley Laboratory 1984).

A number of researchers have suggested developing air-quality sensors and controls as integrated parts of ventilation systems for homes or office buildings (Repace 1982; Lawrence Berkeley Laboratory 1984). For example, a control system could automatically activate an exhaust fan when a gas stove is turned on or a whole-house ventilation system when a pollutant reaches a certain level. In office buildings, the ventilation rates and amounts of makeup air in different zones could be varied on the basis of air quality levels.

F. POLICY ISSUES

Few policy actions have been taken to protect IAQ, especially in residences. This lack of action results from the recent recognition of the hazards, the uncertainties that remain regarding health effects and mitigation techniques, the lack of an IAQ constituency and of an informed public, and the confusion over who is responsible for IAQ and its protection (Sexton and Wesolowski 1984). So far, complaints about "building sickness" have been dealt with in an ad hoc manner by local health authorities.

Some states have gone ahead with IAQ standards or product bans, but such efforts have been piecemeal and uncoordinated (Repace 1982). For example, eight states and additional localities had banned the use of portable kerosene heaters as of Spring 1983 (Meier 1985).

The issue of responsibility must be confronted before public policy initiatives are taken. Given the growing awareness of IAQ hazards and the lack of existing regulations, responsibilities are being determined through private lawsuits in the civil courts. A number of these court decisions are establishing precedents for holding builders, contractors, and product manufacturers liable for injuries and damages resulting from IAQ problems in residences (Everett and Dreher 1982; Anachem, Inc., and Sandia National Laboratories 1982).

Experts conclude that government efforts to safeguard public health related to IAQ should be more extensive and systematic (Sexton and Wesolowski 1984; Spengler and Sexton 1983; Repace 1982). These safeguards would be provided through federal legislation (e.g., as part of the Clean Air Act) and through the rulemaking authorities of agencies like the Consumer Products Safety Commission, HUD, and EPA. Besides instituting product bans or performance standards, governmental entities could require labeling and information programs (Morrill and Geller 1985).

The uncertainties regarding the extent and magnitude of health hazards make it difficult to justify product bans. The controversy over urea-formaldehyde insulation demonstrates the difficulties in this area. The CPSC first banned use of this product in 1982, but the decision was then overturned by the U.S. Court of Appeals. A few states have banned use of this product on their own (Anachem, Inc., and Sandia National Laboratories 1982).

Industry and professional organizations can influence policies and practices through voluntary standards and educational activities. In particular, ASHRAE proposed new ventilation and IAQ standards for buildings in 1981 (American Society of Heating, Refrigerating, and Air Conditioning Engineers 1981). Certain aspects of their recommendations for commercial buildings are controversial and are under review (Janssen 1984). Nonetheless, these recommendations are important because standards of this sort are widely adopted in state and local building codes. Such ventilation guidelines are by no means a blanket solution; additional efforts to reduce pollutant concentrations will be needed where source emission rates are high (Nisson and Dutt 1985). Also, it might be difficult to implement ventilation standards in residences that depend on natural ventilation.

In 1984, South Dakota adopted new energy efficiency standards for homes receiving loans through the State Housing Development Authority. This code allows compliance with either performance or prescriptive standards, and the latter include ventilation and air quality provisions. Homes must meet an air-leakage requirement based on blower-door testing and must include a mechanical ventilation system that provides at least 0.5 ACH (Nisson 1984). The South Dakota regulation is the first residential air quality standard requiring mechanical ventilation in the U.S.

Public awareness regarding IAQ and the ability to change behavior through information, education, and incentive programs will have a bearing on whether regulatory intervention is required. The "publicness" of a particular building may also need to be considered. It may be more appropriate to set standards pertaining to minimum ventilation rates or allowable activities (such as smoking) in commercial buildings than to establish standards for pollutants in residences.

III

**RESEARCH AND PROGRAM DEVELOPMENT:
AGENDAS, THEMES, AND VISIONS**

12

WHERE WE WANT TO GO:
AN OVERVIEW OF PART III

One is reminded of Alice In Wonderland, where Alice asks the cat: "Would you tell me, please, which way I ought to go from here?" To which the cat replies: "That depends a good deal on where you want to get to." Such is the case with developing research needs, including those presented in this book.

A. THE TASK AT HAND

In Part I of this book, we noted that as much as 50% of our current energy use can be conserved through efficiency gains. In Part II, we showed that we have learned how to make new and existing residential and commercial buildings that use very little energy in a cost-effective manner. We also identified programs and policies to realize this potential.

In Part III, we identify several directions in which we need to proceed. First, we need to solve the remaining technical problems that hinder the widespread adoption of known techniques and technologies that are energy efficient. Such problems include the effects on indoor air quality of energy conservation measures, the need for more accurate predictive models of energy use, and the variations observed in savings from identical efficiency measures. Second, we need to improve and apply our knowledge of human and institutional behavior as it affects energy use (e.g., understanding the effects of occupants on energy use and how the markets for appliances and new buildings influence the selection of energy features). Third, we need to work on applications of technology that have been neglected (e.g., multifamily housing and warm-climate building design techniques). Finally, and most important, we need to fully implement the many ideas and techniques that have already been developed; effective implementation requires more experimentation with program-delivery options, greater use of program evaluation, and adequate mechanisms for information and technology transfer.

Such efforts should produce important improvements in the energy efficiency of building envelopes and equipment. However, many new ideas and concepts promise even greater opportunities for energy efficiency. These include "smart" buildings whose walls and conditioning systems are responsive to the external environment and to dynamic fuel prices; building systems responsive to the comfort needs of individual occupants; full-scale energy service companies for the residential sector; and utilities that sell least-cost energy services.

The agenda we present here identifies technologies, building designs, government policies, new delivery services, and information strategies that hold that promise but need testing and development. Some of these ideas will prove unproductive; however, the process of thinking about these needs can uncover other techniques and solutions that have not yet even been imagined. We hope that Part III serves as a stimulus for other researchers, policy makers, and funding organizations.

B. OUR APPROACH

We used a multifaceted process to identify possible short- and long-range research topics and program activities that, in our opinion, should be investigated during the next decade. We drew heavily on our own research, literature reviews, and the thinking that went into preparing Parts I and II. Some of our ideas came from research agendas developed by other organizations.*

We also used approximately 100 visionary ideas prepared and discussed by participants at the 1984 Santa Cruz ACEEE Summer Study. There, each participant was invited to describe his or her "far out" visions of new opportunities to use energy in a more efficient manner. These exercises were called the "One-Minute Visions."

Using the above resources, we compiled lists of research needs, programs to be tested, support services required, and visions of alternate

---

*See other research plans prepared by the Gas Research Institute (1984), Electric Power Research Institute (1984 and 1985), American Institute of Architects (1984), Oak Ridge National Laboratory (1982), U.S. House of Representatives Committee on Science and Technology (1983 and 1984), and the Department of Energy (1984).

futures. Some of the suggestions involve "next steps" for continued refinement and exploration of issues already subject to research. Other ideas identify new lines of inquiry, some with long lead times. We found that some of the research and program development worthy of national attention has a short-term time horizon and some has a long-term time horizon. Our efforts culminated in lists of some 120 near-term and 40 long-term research and program needs. The near-term needs are described in Chapter 13, and the long-term needs in Chapter 14. The needs, both near- and long-term, are listed in Chapter 15.

Rationale for Considering the Visionary Concepts

Only 12 years ago a major report was issued that contemplated the future of energy supply and demand in the U.S. *A Time to Choose: America's Energy Future*, published by Ballinger in 1974, developed three energy scenarios for 1975 to 2000. The Historic-Growth Scenario assumed that U.S. energy use would continue to grow until the end of the century at about 3.4% annually (based on the average rate from 1950 to 1970), leading to use of 187 QBtu per year by 2000. The Technical Fix Scenario assumed similar economic growth but also a determined effort to introduce energy-saving technologies with a resulting energy demand of 124 QBtu per year by 2000. The Zero Energy Growth Scenario envisioned that, through accelerated efficiency improvements as well as redirection of our economy, energy demand would level off at 100 QBtu per year in 2000. To some, this scenario was wildly optimistic about our ability to contain energy growth. One advisory board member commented, "I believe that the changes they (the Technical Fix and Zero Energy Growth scenarios) propose would result in substantial social upheaval, as well as economic stagnation."

In 1984, we actually used the same amount of energy as in 1973 (74 QBtu) while the GNP was 31% greater. In reality, we are consuming far less energy today than was expected or even imagined only ten years ago.

So too, it may be possible to further reduce U.S. energy intensity, both by applying technologies and design practices already available or under development and by pursuing longer-term "visions." We believe new leaps in energy efficiency in buildings may be made by considering the future in open ways, unbiased by institutional, disciplinary, or social or behavioral expectations.

The near-term agenda will help improve energy use in the context of current social institutions. The placement of buildings within the existing fabric of communities is assumed to remain essentially unchanged. Lifestyles, separation of work and living environments, and building-use patterns are expected to continue more or less in their present formats.

This near-term research agenda presented in Chapter 13 is organized around four major topics:

- Integrated approach to building performance;

- Building technologies;

- Institutional and individual influences on energy decisions; and

- Intervention strategies.

For each topic, specific examples of priority research are presented.

The long-term agenda suggests concepts for meeting societal energy needs in different ways from those we know today. This longer time frame approach assumes that greater energy efficiency can be achieved through changes to our social institutions, design practices, and behavioral norms. In the longer-term agenda we look at new ways of meeting the comfort, lighting, and service needs of people within buildings. Many of these longer-term ideas can benefit from careful testing and experimentation beginning now. If proven to have merit, these research activities may allow a planned transition to new forms of energy-efficient buildings, communities, and behavior. This long-term agenda, presented in Chapter 14, includes four topics developed in some detail:

- Alternative energy pricing systems to increase economic and energy efficiency,

- Least-cost energy planning and service delivery,

- People-oriented comfort systems, and

- Dynamic buildings and building components.

Chapter 14 also presents several illustrative ideas from the "One-Minute Visions." By elaborating on a few specific ideas, we hope that readers will think more creatively about our built environment in the 21st century.

We do not set detailed priorities among these research topics nor do we suggest appropriate funding levels for these projects. Our primary aim is not to develop a detailed research program (which in our view is the function of government agencies, industry research organizations, and individual companies). Rather, it is to emphasize the overall need for, and opportunities in, additional research on energy efficiency in buildings.

C. COMMENTARY

ACEEE's Summer Study on Energy Efficiency in Buildings and documents such as this book represent a coming together of ideas from practitioners and researchers sharing a variety of experiences, insights, and expertise. The suggestions for further research and program applications in the following chapters invite a great deal of interdisciplinary effort.

Part II suggests that our understanding of energy use and efficiency has become more integrative. We now recognize technical, economic, and behavioral elements in energy use. We also understand the need to look at the dynamics of whole buildings and not just at the energy requirements of individual components. Similarly, we understand that the wise use of energy is only one of many objectives of successful design. Other concerns include building costs, occupant health and comfort, productivity, and aesthetics. Thus, research environments and program experiments that encourage collaboration among multidisciplinary researchers and professionals are needed. Closer collaboration between the private and public sectors is also important.

13

NEAR-TERM RESEARCH
AND PROGRAM PRIORITIES

A. INTRODUCTION

Much can be done to build upon the huge body of research and program activities conducted during the past decade. First, we need to understand better the interrelationships among energy use and conservation: building-component research, whole-building systems, design and analysis of building environments for nonenergy qualities (e.g., air quality and worker productivity), and the long-term durability of new building designs. This work could provide a comprehensive understanding of "successful" buildings, where success is defined very broadly.

Second, research on ways to apply already developed energy efficiency technologies needs to be increased. Research should focus on markets and target audiences previously overlooked. Examples include energy management control systems for residences and retrofit measures for multifamily residences and mobile homes.

Third, social science research should be applied to energy efficiency. Research on behavior, attitudes, and decision-making processes of individuals and organizations needs to be translated and incorporated into knowledge of how government, utilities, and product manufacturers can influence product purchase and energy-use behavior.

Fourth, the results of past research must be put to good use. Research findings and experimental program results need to be translated into relevant policies and programs. They, in turn, need to be applied through building construction, appliance and equipment sales, and program intervention strategies. The effectiveness of these applications should be communicated back to the researchers, providing a feedback loop to better focus future research.

The near-term priorities listed here are organized around four themes: an integrated approach to building performance; building technologies; institutional and individual influences on energy decisions; and intervention strategies (purposeful efforts to alter the normal operations of the marketplace).* While many of the topics are not new, they are presented in a way that we hope conveys the breadth, complexity, and interrelatedness of building energy research. Chapter 15 presents our list of specific research and program suggestions from which these priority topics are drawn.

B. INTEGRATED APPROACH
TO BUILDING PERFORMANCE

Whole-building research recognizes the importance of treating building elements, occupants, and the building surroundings together to understand the overall operation of a building. Quantitative analyses can examine relationships among energy use, indoor air quality, and worker productivity. Other aspects (such as thermal comfort, lighting, smell, sight, acoustics, and occupant satisfaction) are also important dimensions of whole-building performance. Energy efficiency should complement and not compete with these qualitative characteristics. It is important to integrate energy efficiency into the overall building design and operation process. That is, we should search for synergisms between energy efficiency and other building attributes (e.g., we should identify low-energy lighting systems that are aesthetically pleasing to building occupants).

Priority research is described for four topics: whole-building energy performance, long-term monitoring of building performance, environmental qualities affected by energy-efficiency actions, and low-energy buildings in hot climates.

WHOLE-BUILDING ENERGY PERFORMANCE

Description of the Research Area

Past energy efficiency research emphasized improvements in building components, equipment, subsystems, and materials. The focus on building "parts" needs to be complemented with a greater understanding of how the

---

*Although the topics presented here cover a wide range of important issues, other possible areas are not mentioned. These include community energy systems (such as district heating), community energy planning, heat pumps, advanced lamp technologies, and laboratory equipment and procedures to measure the energy characteristics of building components.

parts affect overall building energy efficiency. Individual parts may be highly energy efficient but, when placed within the context of a building, may not yield an effective whole. For example, the addition of insulation may reduce heating costs, but may also increase cooling loads, because internal heat loads can no longer be transferred through the envelope to the exterior.

Designers of buildings need to assess energy saving options and make trade-offs among building features to optimize overall energy performance. The objective is to make building components work synergistically. To achieve this optimization, more information is needed on interactions among building elements, between building elements and occupants, and between buildings and their surroundings. This need is especially critical for commercial buildings and multifamily residences; more attention is also needed on buildings in hot-dry and hot-humid climates.

One of the strongest arguments for this type of research is the consistent discrepancy between simulated energy performance and actual performance. Such discrepancies may occur because building subsystems perform differently in isolation than when they are integrated into the total building.

Issues to Address

Research needs to focus on many areas. Whole-building energy related indices with appropriate definitions, uses, and measurement techniques need to be developed for a variety of building types. Such indices would include various types of Btu/ft^2-yr consumption classifications; load profiles; performance integrity and degradation factors; average U values for building components; infiltration levels; thermal capacitance; surface-to-volume ratios; percent glass; solar-load ratios; lighting efficiencies including daylighting; comfort; and indoor air quality. These indices should support development of energy performance guidelines and standards (e.g., ASHRAE) and of methods to assess the effectiveness of building retrofits.

Uniform standards for the collection of data related to whole-building energy performance, simulation, and documentation need to be further developed (building on work conducted by SERI, ASHRAE, and others) to allow comparisons and to meet the needs of different building-industry sectors.

Analytical tools for diagnosis, design, and long-term tracking need to be developed to establish and maintain data bases on whole-building energy performance. This area also includes identifying and measuring critical climatic and context data. An example would be a tool that not only records whole-building energy performance but also analyzes building subsystems and occupancy patterns.

Another important question concerns the sensitivity of whole-building energy performance to major design, construction, and retrofit decisions. Analyses need to compare simulated performance at various times during building design and construction with actual whole-building energy performance. Such research should seek to understand how decisions on building shell, HVAC equipment, and associated construction methods and quality control affect overall building performance. In addition, the impacts of major architectural features (e.g., atria and earth integrated structures) on energy performance need to be analyzed.

The effects of construction processes and methods on whole-building energy performance also need to be investigated. Different processes, delivery systems, methods of construction, and quality controls may produce differences in energy performance.

The sensitivities of whole-building performance to major interior design characteristics need to be described. Interior design elements (e.g., partitioning, "office landscapes," convective air exchange systems, thermal and illumination zoning, and location of thermal mass) affect energy use in ways that are poorly understood.

The impacts of occupancy patterns on building performance need to be better understood. Relevant variables include sensory quality expectations, building ownership, who the people are, what they do, when they are in the building, who pays for fuels, and occupant turnover.

Research is needed to determine how responsibility for building performance might best be shared among the architect, building contractor, owner, manager, and occupants. The importance of government thermal performance standards relative to industry norms is a related issue.

Possible Approaches

Methods of designing and monitoring building performance need to be expanded. To do so, new guidelines and improved instrumentation for performance monitoring should be developed, that include (1) appropriate time periods for data collection and (2) the relationships between energy consumption and other attributes of building performance. Data requirements need to be defined for these nonenergy building attributes, and the BECA data bases developed by LBL should be expanded to include these nonenergy factors. In addition, BECA-type data bases should include comparisons of actual and simulated performance of new and retrofit buildings.

Participation of social scientists is critical in developing research methods as well as in interpreting results. Engineers, architects, and social scientists should work together more closely to study occupant responses to low-energy buildings. New measurement and diagnostic tools may be needed for automated sensing, control, recording, and analysis. Such automated systems can reduce data collection costs and ensure the validity of long-term tracking of buildings.

Related work

Past efforts in this area were described in Part II. Other reports on whole building energy performance are available from BTECC, ASHRAE, AIA Foundation, and DOE. BTECC has a Research Coordinating Council devoted to whole-building thermal performance. The BECA data base and related research carried out at LBL provide significant whole-building data for analysis and interpretation.

LONG-TERM MONITORING OF BUILDING PERFORMANCE

Description of the Research Area

As discussed in Chapters 4 to 7, considerable data exist on the performance of low-energy buildings and building equipment. In a few cases, related issues (such as occupant satisfaction or product degradation) have been studied. The DOE Passive Solar Commercial Demonstration Program is a good example of such analysis. However, these energy conserving buildings, equipment, and technologies are quite new. As a result, performance data generally correspond to initial periods of use. It remains to be seen how well highly efficient buildings and equipment perform over an extended time.

Issues to Address

Many questions about long-term performance of buildings and their components need to be investigated. For example: Will the vapor barriers in superinsulated homes remain intact for 20, 30, or more years? How will moisture affect well-insulated, tight homes over an extended period? Are new heat-pump water heaters reliable? Will they last as long as (or longer than) electric-resistance water heaters? Can well-designed commercial buildings with sophisticated control systems be operated effectively over the long term? Will commercial buildings with energy efficient lighting and HVAC systems accomodate changes in occupancy patterns without sacrificing energy efficiency or occupant satisfaction? How are the economics of retrofit and new low-energy buildings affected by the lifetimes of conservation measures? How do these lifetimes vary as functions of climate, building type, and maintenance procedures? What additional research on techniques or standards could be undertaken to increase equipment and system lifetimes and operating reliability?

Answers to these questions are needed to judge the technical, economic, and practical value of various conservation measures. Savings degradation over time and the absolute lifetime of the various measures are crucial factors in building, equipment, and program assessments. At the present time, we can only guess at these factors.

Possible Approaches

This research involves systematic monitoring of selected buildings over an extended time to determine energy performance, maintenance requirements, lifetime usage characteristics, occupant satisfaction, and other factors related to low-energy buildings and building equipment. Research costs can be kept down by using records maintained by building owners and managers supplemented by periodic specialized surveys and independent measurements of building performance. A representative and statistically significant sample of buildings should be followed for 10 years or more. As new, potentially important conservation techniques begin to be adopted on a wide scale, additional buildings should be added to the sample.

It may be possible to bring together a broad range of organizations interested in such data. The DOE, AIA, NAHB, and NIBS might jointly sponsor long-term monitoring of high energy efficiency houses; DOE, GRI, and relevant trade associations might sponsor long-term monitoring of highly efficient gas furnaces and gas heat pumps; and DOE and EPRI might sponsor long-term monitoring of heat pump water heaters and other advanced electric HVAC systems and appliances.

Related Work

DOE sponsors the collection and evaluation of energy performance data on new and existing residential and commercial buildings, but makes no special effort to follow the same buildings over an extended time. Utilities and some research organizations have occasionally collected long-term performance data. For example, Alabama Power has been collecting and evaluating data on the maintenance requirements and lifetime of heat pumps for many years.

ENVIRONMENTAL QUALITIES AFFECTED BY ENERGY EFFICIENCY ACTIONS

Description of the Research Area

Environmental qualities include health, comfort, emotional well-being, and human productivity. Energy conservation strategies can enhance or diminish environmental qualities. Research and development on energy conservation and indoor air quality is underway, but little attention has been paid to the overall impact of energy conservation strategies on environmental qualities.

Issues to Address

Air quality

When the amount of fresh air entering and leaving a building is reduced, the level of pollutants and odors in the air may rise. The nature of that degradation and the effects it has on building occupants need to be better understood. Strategies for controlling indoor air pollutants (including tobacco smoke) at their source need to be further developed. Also, mechanical ventilation systems for residences need to be improved and commercialized. Finally, mechanisms to ensure adequate IAQ in buildings need to be developed, studied, and implemented.

Research is needed to examine the role of regulations in maintaining adequate levels of indoor air quality. Should smoking be prohibited or limited in public buildings? Should the use of certain building-component materials be restricted or banned? Should enforcement be triggered by complaints only, be based on random samples, or be the responsibility of building owners and managers?

Thermal comfort

How comfortable we feel is not always reflected by the thermostat setting. Comfort is determined also by humidity, air flow rates, mean radiant temperatures of interior surfaces, activity levels and their nature, homeostatic processes, and personal expectations. Research is needed on occupant comfort in low energy buildings, especially those with passive solar designs or sophisticated control systems.

Luminous qualities

Daylighting and other passive energy systems can save energy but can also cause irritation by producing glare, reflections, and stark contrasts.

Little data exist on the luminous qualities of indoor spaces. Research is especially needed to assess the light characteristics associated with alternative daylighting strategies, including those that are integrated with passive solar heating and/or cooling systems. Daylighting computer models could be improved to better predict lighting levels in various interior spaces as functions of time-of-day, season, and cloud cover.

Aesthetics

Some objects are pleasant to look at, while others cause unpleasant reactions. The clutter of rooftop solar collectors, added without concern for appearance, is ugly to some people. Efforts are needed to integrate the design of new energy efficient buildings with their overall visual appeal.

Productivity

Worker salaries are roughly 100 times greater than energy costs in commercial buildings (Chapter 7). For these buildings, the impact of energy efficiency improvements on worker efficiency, comfort, and satisfaction is very important. Research is needed on the relationship between worker performance in office and other commercial buildings relative to lighting types and levels, ventilation rates, indoor temperatures, and other energy related factors.

Possible Approaches

Social scientists, psychologists, and economists need to join with engineers, architects, and others to assess the impacts of energy efficient technologies and their applications. Case studies of low energy buildings should include explicit attention to these environmental factors. This may require new methods to measure and quantify these factors.

The fields of social science, psychology, and the arts must be encouraged to provide measures of the sensory qualities of the built environment. We need to express and make manifest these factors in the design process. This approach may require science and engineering-oriented institutions to fund new types of research.

Related Work

The AIA maintains an extensive library related to the aesthetics of buildings; ASHRAE is active in areas related to air quality, ventilation, temperatures, and humidity levels in buildings; the Illuminating Engineering Society focuses on artificial and natural lighting in buildings; and DOE's Passive Solar Commercial Buildings Program examined a broad range of issues related to low energy buildings.

LOW-ENERGY BUILDINGS IN HOT CLIMATES

Description of the Research Area

Although new construction is concentrated in hot-climate areas of the U.S., relatively little research has focused on energy conservation problems for these climates. In such areas, air conditioning consumes large amounts of energy and can be the primary contributor to electric utility peak loads. Under these conditions, different approaches might be preferred for conserving energy than are preferred in colder regions. Less emphasis might be placed on improving thermal integrity of building envelopes, and more emphasis placed on reducing internal heat gains in buildings. Results of such research should be helpful in improving the thermal performance standards for new buildings and in designing better retrofit programs for existing buildings.

Issues to Address

Several hot-climate topics need further investigation. First, the effect of improving building thermal integrity on space conditioning requirements (air conditioning as well as heating) should be studied more vigorously. Field evaluations of the impacts of retrofitting buildings in hot climates should be conducted and should consider peak-power as well as energy savings.

Second, the potential for making basic improvements in air conditioners and heat pumps as well as for developing alternative cooling techniques should be further explored. Promising technical options for compressor-driven systems include the use of new refrigerant mixtures and variable-speed operation. Also, the development of high efficiency air conditioning systems with high dehumidification capabilities should be continued; heat pipes and desiccants both appear promising for efficient dehumidification. Finally, indirect evaporative coolers could greatly reduce energy use in many locations.

Third, many building technologies, systems, and design features now have the potential for reducing cooling requirements. Research is needed to determine the cost and performance of these alternatives. It may be possible to develop window materials or coatings that can vary the amount of heat and light transmitted. Radiant barriers cut down on heat transfer across air spaces via radiation. Thermal mass within the building itself or independent systems for cool storage can be used for energy conservation as well as load management.

Fourth, moisture control in low-energy homes in hot, humid climates is a major problem. Some guidelines related to moisture problems in hot climates are available, but much remains to be learned about such fundamental issues as moisture transport in buildings and moisture control in attics.

Fifth, air-conditioning requirements in commercial buildings are of primary concern because of the combination of thermal and internal gains. Consequently, research should focus on ways to reduce air-conditioning energy and peak-power demands with thermal mass, cool storage, more-efficient lighting systems, and innovative control systems. The relationship between lighting and cooling loads, especially in large commercial buildings, is a critical concern. Additional research is needed on the use of thermal storage and passive cooling techniques (convection, evaporation, radiation, and earth-cooling) for commercial buildings.

Sixth, the relationships between air-conditioning and heating systems need further attention. Optimizing HVAC equipment, window size and orientation, and insulation levels for air conditioning use may not yield desirable space heating performance. Preferred combinations that provide satisfactory performance and costs in both heating and cooling seasons will vary with region and fuel price.

The ability of building simulation models to predict energy and power consumption in hot climates is highly uncertain, in part because of uncertainty over occupant management of windows and air conditioners. Additional data is required on occupant behavior, partial load performance of HVAC equipment, internal gains by time of day, and characteristics of solar apertures. Better data in these areas may yield substantial improvements in the predictive power of these building simulation models. This need for better data is of particular concern when newer energy conserving features, such as radiant barriers or lighting control systems, are added to buildings.

Possible Approaches

For technologies and design approaches at an early stage of development (e.g., dynamic windows, variable-speed air conditioners, use of phase-change materials for cool storage, and desiccant air conditioners), laboratory R&D as well as prototype demonstration and monitoring are required.

Several energy conservation techniques for buildings in hot climates have recently or will soon become commercially available (e.g., heat-pipe-assisted air conditioners, thermal mass for cool storage, and advanced building control systems). Field studies should be carried out to investigate integration of these new technologies or designs into buildings, methods of operation, and overall building performance. Field performance studies should examine indoor conditions, occupant comfort, and operator behavior relative to energy consumption. The monitoring and analytical techniques developed primarily for climates where space heating dominates should be tested for their usefulness in hot climates.

Building simulation models should be refined to accommodate new energy-conserving features, and should be validated against actual building performance. Also, since peak load and load management are of concern, models should account for both electric demand and energy requirements.

Studies of the cost effectiveness of the emerging technologies are also needed. Data bases on the energy and economic performance of conservation techniques for both new and retrofit residential and commercial buildings in hot climates could be established. Once performance is well understood, methods for estimating the cost-effectiveness of conservation measures as a function of key variables (such as climate, building type, building usage, and utility load shape and marginal costs) could be developed.

Related Work

DOE sponsors some research and development related to conservation in hot climates, including projects on building materials, advanced windows, heat-pump technology, and the monitoring of residential buildings that have innovative conservation features. GRI has a program on energy efficient gas fired cooling systems, and EPRI supports work on electric heat pumps and load management.

The Florida Solar Energy Center is developing equipment and performing studies related to air conditioning in hot, humid climates. Much of this work is aimed at conserving energy through the use of radiant barriers, air conditioners with high latent cooling capacities, and passive cooling. Prototype advanced air conditioners that use heat pipes and desiccants have been developed.

Equipment manufacturers conduct research and development on a variety of techniques and products for reducing energy use in hot climates.

C. BUILDING TECHNOLOGIES

A wide variety of R&D is needed on building technologies in relation to energy use and conservation. First, how specific conservation measures actually perform in comparison to engineering calculations and how occupancy patterns, building use, and climate affect performance need to be investigated. Second, new and improved technologies are needed for building control systems, building materials, appliances, and integrated energy systems.

For example, whether the monitoring and control technologies developed for large commercial buildings can be used in smaller commercial facilities and in residences remains to be seen. Technologies that need to be developed, especially for small commercial and residential applications, include user-programmable controls and information-feedback mechanisms. In developing these technologies, how users and operators interact with monitoring and control systems, how they use control equipment, how they respond to various situations with monitoring equipment, and how they get information from the equipment must be understood.

New or modified retrofit techniques need to be developed for single-family, multifamily, and mobile homes. For these and commercial buildings, waste-heat-recovery systems, cogeneration facilities, and thermal-energy-storage systems need to be developed and applied. We need to know how these conservation measures actually perform in comparison to engineering calculations and in relationship to the performance of alternative conservation measures.

Finally, the energy efficiency of residential and commercial appliances still leaves substantial room for improvement. Appliances, air conditioners, heating equipment, and lighting products are all important building technologies that merit further R&D. However, much progress has been made in improving the efficiency of these technologies during the past decade (Chapter 6), and new energy efficient products reach the marketplace almost every month. Because of the rapid advances that are occurring and the substantial R&D activity in the private sector, we do not address R&D needs in these areas.

BUILDING CONTROL TECHNOLOGIES

Description of the Research Area

Building control technologies (computers, sensors, and controls) have been developed over the past decade primarily for use in large commercial buildings. More recently, home control systems have been commercialized (Fig. 13.1). They control lights, appliances, fire, and security systems and they allow computerized communication.

Additional R&D is needed on control technologies for residential and small commercial facilities. Products that create "smart" or "self-correcting" buildings through integrated monitoring, feedback, and control strategies would be useful. Such technology should include the ability to adapt to changes in season, outside temperature and humidity, building usage levels, and utility rates (real-time, time-of-day, or other signals).

Such controls could automatically operate heating and cooling equipment, appliances, and adjustable building elements (e.g., window shades and ventilation louvers) to minimize fuel costs while maintaining occupant comfort. Research is needed to determine the energy and electric load reductions produced by computerized controls by building type, control system type, and climate.

For commercial buildings, a key research need is to understand how building operators use and respond to building monitoring and control equipment, and whether information is presented to assist operator understanding of energy system inefficiencies. Similarly, for households, actual savings will likely depend on how residents use control equipment.

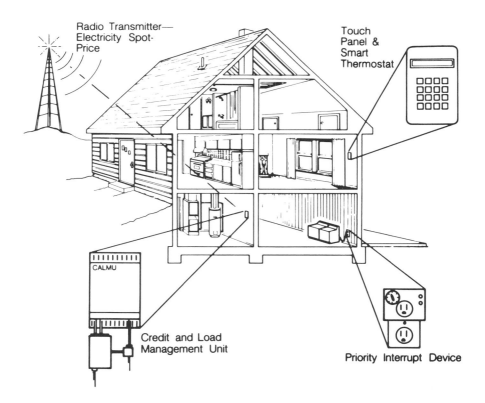

Fig. 13.1. The British Credit and Load Management System. CALMS is an example of recent advances in telecommunications and microprocessors, which can provide improved monitoring and control of residential and commercial buildings.

Issues to Address

Should the controls be permanent or portable? Should their power source be through power line carriers, dedicated lines, or phone line carriers? How susceptible are these technologies to errors in construction and installation; proper commissioning, testing, and verification; and training of operating personnel?

Can low-cost sensors and control actuators be developed and produced to ensure greater market penetration?

Will provision of more energy use information to building operators lead to more efficient building operation? And what information displays and user interactions might facilitate greater energy and cost savings? Alternatively, can these control systems reduce the role of building operators and thereby overcome traditional problems associated with poorly trained, paid, and motivated building operations staff? Will people enjoy living and working in smart buildings?

Possible Approaches

Field studies under controlled and real-world conditions are needed to evaluate user response to control systems and actual energy savings. A variety of control systems and operating procedures should be tested in different building types. For example, data on energy use, peak loads, operating costs, and occupancy patterns could be collected before and after installation of control systems. Also, control and feedback systems could be tested to see if they substantially affect building and electric utility system peak loads. These studies might include some utility control of building loads (to provide the utility with dispatchable load management).

The tradeoffs among control system sophistication, cost, ease of use, and energy or load savings should be assessed for different systems and building types. For example, a simple timer might sometimes be more effective than a complicated computerized energy management system.

Control systems consist of software as well as hardware. The former affects system operation and provides many opportunities for innovation and improvement. For example, a program might be developed to use weather forecasts (transmitted periodically to the control system) to anticipate energy demand in the building, and to precool or preheat the building to increase efficiency and reduce peak power demand.

Related Work

Papers by Novey, Warren, and Peddie from the 1984 ACEEE Santa Cruz conference and Chapter 8 of this book treat related issues. Work on the Smart House by the NAHB Research Foundation and research on building automation conducted by Tishman Research Corp. are also relevant, as is the British experience with their residential Credit and Load Management System (CALMS Fig. 13.1). Finally, Sorenson (1985) discusses the role of feedback on household energy use.

BUILDING RETROFIT TECHNIQUES

Description of the Research Area

Better data are needed on actual performance of retrofit technologies and techniques to understand how various factors affect energy savings (Fig. 13.2). Such information could be useful to several groups. Managers of low-income weatherization programs and private sector retrofit companies need better information to provide more-optimal retrofits. Energy auditors could provide more reliable recommendations to building owners and managers. Energy planners and analysts could make better assessments of the conservation potential in buildings. Improved data on the costs and performance of retrofit measures, and the factors that affect

RETROFIT ENERGY SAVINGS DEPEND ON:

✔ ENERGY AUDITOR	✔ QUALITY OF MATERIALS
✔ AUDIT CALCULATIONS	✔ QUALITY OF INSULATION
✔ BUILDING CHARACTERISTICS	✔ OCCUPANT BEHAVIOR
✔ MEASURES SELECTED	✔ OWNER MOTIVATION

Fig. 13.2. More research is needed on the factors that affect retrofit performance, especially in multifamily and commercial buildings.

performance and cost, would be enormously helpful in improving several major federal programs (e.g., the Residential Conservation Service, Commercial and Apartment Conservation Service, Institutional Conservation Program, and low-income Weatherization Assistance Program). Such data would be particularly valuable for multifamily and commercial buildings.

Issues to Address

In the residential sector, additional research is needed to understand the variation in cost-effectiveness among retrofits and the ways to minimize these variations. Savings can be affected by a wide range of factors, including type of measure, quality of installation, characteristics of the building, and behavior of the occupants. Research is needed to understand the generally poor agreement between savings predicted in an audit and actual savings after the retrofit is carried out and to improve the accuracy of audit predictions. Also, it would be useful to update conservation supply curves on the basis of actual performance results and provide curves for a wider set of housing types and climates.

Retrofits of multifamily buildings and mobile homes have received little attention to date. For these building types, what are the most promising retrofit measures; quantitatively, what factors cause savings (or lack of savings); and how can the savings and cost effectiveness of multifamily and mobile home retrofits be maximized?

In the commercial sector, how can the cost, savings, and related factors for individual retrofit measures in different building types (e.g., lighting system retrofits in schools, boiler retrofits in hospitals, and HVAC retrofits in high-rise office buildings) be determined? Can conservation supply curves be developed for the various commercial building types based on the cost and savings results from specific retrofit projects? Can these supply curves be expanded to include the costs of conservation programs and to reflect delays associated with program implementation and consumer behavior? Finally, what are the lifetimes of retrofit measures and the persistence of energy savings?

Possible Approaches

To address variations in savings among retrofits in single-family homes, many homes could be retrofitted under closely monitored conditions. Detailed data could be collected on preprogram and postprogram energy consumption by end use, microclimate conditions, the characteristics of the house, the measures installed, the quality of the installation, and occupant behavior. Changes in occupant behavior (e.g., indoor temperatures or wood heating) caused by retrofit may substantially affect energy savings. Careful analysis of the data might show the importance of these factors on actual energy savings and might indicate whether and how simulation methods can be improved. The results might also suggest ways to better target homes for retrofit treatment and the actions to be taken to achieve a larger fraction of the potential savings identified in these homes.

To determine optimum retrofit strategies for multifamily buildings, mobile homes, and commercial buildings, similar experiments could be conducted. With commercial buildings, particular attention should be paid to building operation and maintenance, occupancy, and weather.

To address the issues of degradation of retrofit measures and energy savings, the durability of products and materials (such as insulation, caulking, weatherstripping, window films, and HVAC equipment) could be studied under simulated and real-world conditions.

Related Work

Current knowledge regarding retrofit measures, their savings, and the relationship between predicted and actual performance of retrofits for single-family homes is well documented in the proceedings from ACEEE Santa Cruz summer studies. DOE supports an ongoing building retrofit research program with projects on single-family, multifamily, and commercial buildings. The DOE program currently emphasizes mechanical equipment retrofits for low-income housing, discrepancies between predicted and actual savings in housing retrofits, strategies for retrofits in public housing, and the collection of data on commercial building retrofit measures.

D. INSTITUTIONAL AND INDIVIDUAL INFLUENCES ON ENERGY DECISIONS

Government managers, utility managers, manufacturers, distributors, and installers of energy efficiency products and materials need to understand the decision-making behaviors of individual energy consumers and energy-using organizations. This objective can be accomplished by extending the body of social, psychological, and anthropological research on decision-making and consumer behavior. Such information could guide efforts to influence decisions toward outcomes with increased energy efficiency.

Behavioral research and the complementary analysis of prior conservation programs should be used to design future programs, select marketing and delivery techniques, and determine the effective content of promotional media.

To promote energy efficiency, government agencies may develop incentives, regulations, or other mechanisms to bolster utilities' and building industry organizations' activities. Special attention should be given to incentive mechanisms, such as raising utility profits for aggressive energy conservation programs or offering payments to builders who exceed minimum building energy efficiency levels.

APPLICATIONS OF ENERGY BEHAVIOR RESEARCH

Description of the Research Area

Substantial research has been conducted into consumer behavior and energy decision processes. Applications of this research, however, are critically needed to increase actions leading to more efficient use of energy.

Decisions with energy efficiency implications are made in at least three stages: building design, equipment specification and selection, and building operation. These decisions may be made by different individuals or organizations, including owners, occupants, manufacturers, building designers, or intermediaries. Each is driven by different motivations in selecting and using energy efficient products.

The contributing factors to individual and organizational decisions have been modeled in various ways (Chapter 8). Two that may be particularly relevant are the social-diffusion and problem-avoider models. The social-diffusion model considers behavior to be influenced by social situations, communication networks, and relationships to others in the network. The problem-avoider model assumes that behavior is motivated primarily by dramatic changes in the physical or social environment and that action is (only) taken to prevent the occurrence of negative outcomes. Development of future energy policies and programs should build upon these two models.

Experiments and demonstrations are needed for such topics as motivating actions for energy efficiency, effective communication techniques, credibility of the information and its source, eliminating barriers to action, and gaining support of critical intermediaries. The experiments and demonstrations should result in guidelines to policy makers, program operators, and energy service businesses on how to gage customer acceptance of programs and services, determine the degree of personalization and control required, specify clear marketing objectives, target audiences likely to be receptive to the product or service, design and deliver programs for maximum market penetration, and select appropriate promotional and information-distribution strategies.

Issues to Address

Can low-cost experiments be designed and implemented to yield timely and relevant information about consumer behavior? Can programs or target audiences be identified whose potential for energy efficiency is large enough to warrant experimental investigation? Can different types of individuals and organizations be classified according to the decision-making models suggested above? Can behavioral models and research explain decisions and tradeoffs among the multiple dimensions of energy-related decisions? Can conservation program effectiveness be improved by basing such programs on these models of consumer behavior?

Possible Approaches

Prototype models of decision-makers could be developed by market segment (e.g., single-family home owner or commercial-building tenant); type of investment (e.g., new building or equipment, retrofit, or operation and maintenance); psychological profile (e.g., risk takers, early adoptors, or pack followers); social or cultural values (e.g., glamour, aesthetics, or independence); program-intervention type (e.g., information, incentives, technical assistance, or service delivery); and other appropriate characteristics. These models could serve as standards against which policies, programs, and market delivery techniques can be evaluated for their effectiveness.

Conservation marketing techniques could be developed to differentiate among market segments, conservation action-type, and/or psychological profile of the decision-maker.

The use of "trade allies" (e.g., appliance dealers, neighborhood groups, local realtors, and builders) should be further explored.

Related Work

Papers by Archer et al., Diamond, Ruderman et al., Wilk and Wilhite, and Norton from the 1984 ACEEE Santa Cruz conference discuss consumer behavior. See also Arbor (1984), Booz Allen & Hamilton (1985), Clinton (1985), Morrison and Kempton (1984), Stern and Aronson (1984), and Williams (1983) for related references.

E. INTERVENTION STRATEGIES

Intervention strategies are policies, programs, incentives, or regulations intended to influence market and behavioral decisions related to energy use. Careful planning and design of implementation strategies are necessary if these interventions are to be effective. Implementation should be based on experimental program designs, careful selection of program alternatives, and documentation of program results. Implemented programs must be validated to verify if the outcomes were those either anticipated or desired.

A critical need in such evaluation is the development of analysis tools that program managers and practitioners can employ themselves. Also, guidelines are needed for the selection of evaluation methods appropriate to the program being evaluated, for the choice of valid sample designs, and for related analytical issues. Whether or not programs are evaluated may also depend on the requirements of utility regulators and/or governmental funding agencies.

The transfer of planning and evaluative research results to professionals and program managers should be systematic and in forms structured for ready use. Packaged guidelines could address steps in implementation, outreach techniques, the content of promotional material, and effective ways to assess program results.

Professionals, practitioners, and program managers (those on the front lines of "energy efficiency duty") must have the logistical support they need. They must also have the opportunity to give researchers feedback on how useful the research results were, what refinements in research investigations are needed, and what additional areas of research or analysis could support operational activities. Appropriate communication mechanisms are needed to sustain this feedback.

INCREASED USE OF CONSERVATION PROGRAM EXPERIMENTS

Description of the Research Area

During the 1970s and early 1980s, government agencies, utilities, and private firms conducted an incredible variety of energy conservation programs. These programs provide, in principle, an enormous library of experience on which to draw in designing and implementing new and improved programs.

Building upon the experience of past programs is difficult for three reasons: haphazard selection of program alternatives, inadequate or nonexistent documentation, and insufficient use of experimental designs that permit reliable interpretation of program results.

We can learn much more from ongoing and planned programs if they are designed to yield information that is reliable and generalizable. Unfortunately, most programs are planned, designed, and operated with insufficient inputs from social scientists and statisticians knowledgeable about research methods. The main thrust of pilot programs seems to be on testing program delivery mechanisms, ensuring that the program can be operated in the field as intended. Additional planning, however, could make these pilot programs much more informative. Careful attention to the critical program elements to be tested, to selection of areas in which the program will be offered, to selection of customers to be offered the program or placed in a control group, to design of data collection instruments, and to subsequent analysis of the data could pay off handsomely (Fig. 13.3).

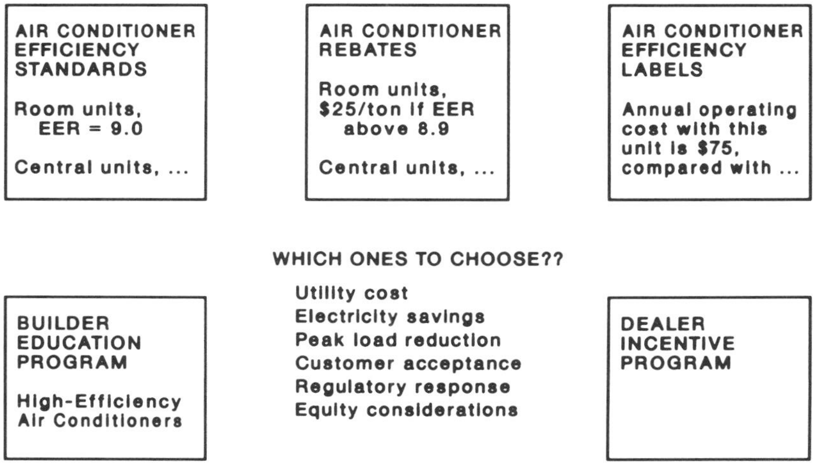

```
┌─────────────────────┐   ┌─────────────────────┐   ┌─────────────────────┐
│ AIR CONDITIONER     │   │ AIR CONDITIONER     │   │ AIR CONDITIONER     │
│ EFFICIENCY          │   │ REBATES             │   │ EFFICIENCY          │
│ STANDARDS           │   │                     │   │ LABELS              │
│                     │   │ Room units,         │   │                     │
│ Room units,         │   │  $25/ton if EER     │   │ Annual operating    │
│  EER = 9.0          │   │   above 8.9         │   │ cost with this      │
│                     │   │                     │   │ unit is $75,        │
│ Central units, ...  │   │ Central units, ...  │   │ compared with ...   │
└─────────────────────┘   └─────────────────────┘   └─────────────────────┘

                     WHICH ONES TO CHOOSE??

                     Utility cost
┌─────────────────────┐   Electricity savings     ┌─────────────────────┐
│ BUILDER             │   Peak load reduction      │ DEALER              │
│ EDUCATION           │   Customer acceptance      │ INCENTIVE           │
│ PROGRAM             │   Regulatory response      │ PROGRAM             │
│                     │   Equity considerations    │                     │
│ High-Efficiency     │                            │                     │
│ Air Conditioners    │                            │                     │
└─────────────────────┘                            └─────────────────────┘
```

Fig. 13.3. Small experiments and pilot programs would provide valuable empirical information on optimal ways to improve energy efficiency through government and utility conservation programs. Each box includes a different conservation program alternative. The unboxed list in the center identifies criteria that might be used to judge the performance of these options, singly and in combination.

Issues to Address

A key issue is whether experimental designs can be developed and applied that will incorporate careful selection of program alternatives, random assignment of participants and nonparticipants, and use of a control group. These elements would ensure that results obtained from the experiment are relevant and would be reasonably generalizable. The variables to examine with such designs should include the costs of the different approaches, the response rate to each approach, the relationship between response rate and participant characteristics, and actual energy and load reductions.

Possible Approaches

Experiments could be conducted to compare the effectiveness of various marketing media (telephone calls vs letters), sources (the utility employee who audited that house, the utility as a corporation, the contractor who would install the measures, or a local community group), and messages (save money, save energy, increase comfort, or fulfill a social responsibility).

The costs of implementing various marketing approaches could be tracked and correlated with the success of each approach in enrolling customers in energy efficiency programs. Follow-up surveys might be conducted to determine why subjects did or did not respond and to identify demographic characteristics (age of household head or income) and attitudes that might have affected responsiveness to different approaches.

Similar experiments could be designed to test the effects of different program options, such as types and levels of financial incentives (e.g., cash rebates vs low-interest loans) and different energy auditing techniques (e.g., detailed engineering audits, house-doctor audits, and simple walk-through audits).

Related Work

The general evaluation literature includes many discussions of design and implementation of program experiments. Unfortunately, these principles have only rarely been employed in utility and government conservation programs. The proceedings of two recent conferences on utility conservation programs include several related papers; see Synergic Resources Corp. (1984) and Pacific Gas & Electric Company (1985).

INFORMATION AND TECHNOLOGY TRANSFER

Description of the Research Area

For more than 10 years, research has been conducted on the energy efficiency of building components, materials, and systems. Significant experimentation and monitoring of the performance of energy efficiency measures have also been conducted. This work has been carried out by national laboratories, building industry research groups, and manufacturers. Much useful information has been obtained, but little has been effectively presented to building industry practitioners.

The building industry, contractor associations, professional societies, state and local governments, and others disseminate energy information through their publications, conferences, and other means. Few of these media present research results in a form useful to manufacturers and field practitioners.

What is missing is a critical information transfer step between research and commercialization accompanied by market acceptance. In the building industry this problem explains the 10- to 20-year time lag between research completion and commercial acceptance. Mechanisms, formats, and distribution channels to carry out information and technology transfer need to be developed.

Issues to Address

What technology transfer roles can be played by the research community and the building industry, and what potential exists for collaborative efforts?

What specific constituencies in the building industry need to be reached, and what needs must be met before they can adopt energy innovations (e.g., information, resources, or incentives)? What types of information (e.g., technologies, products, practices and methods, systems integration, performance information, and user feedback) need to be transferred within the building industry?

Who will ensure the quality of the information submitted to clearinghouses, information centers, or building industry publications, and what criteria will they use? How will the findings from building demonstrations or other field evaluations be verified?

Can a system in which information and data about energy efficiency is maintained and accessed be created, and at what cost?

How can transferable findings of research efforts be isolated, stored, retrieved, and maintained in a timely manner? How can such a system of information transfer be funded?

Is adoption of research innovations by manufacturers hindered by patent and licensing issues?

Possible Approaches

A research clearinghouse could be established for energy-conservation program managers, government energy analysts, and building industry experts. Such a clearinghouse should contain high quality information on building energy research of three types:

- Technical information on materials, systems, and other building components.

- Programmatic information on ways to convey technical knowledge to contractors, building industry suppliers, and architects and engineers. This information should include the results of program evaluations and demonstration programs.

- Supporting information on methods to rapidly introduce energy efficiency methods into routine building practice.

National, regional, and local forums could be arranged for the exchange of applied research findings between researchers and building industry opinion leaders. This exchange might be accomplished through:

- A journal on energy-efficient building applications. *Energy and Buildings* is a technical journal, aimed primarily at researchers. *Energy Auditor & Retrofitter*, on the other hand, translates research results into a format understandable to practitioners. The need for other journals similar to EA&R should be assessed.

- Publications and presentations to transfer the results of research projects to industry. Such research-transfer activities might be carried out by communications specialists rather than the researchers themselves.

Application materials could be packaged in forms directly usable by building industry professionals: building owners and developers, contractors, commercial building managers, energy conservation program managers, and architects. The material should communicate results of research efforts on a given subject. Format might include design tools, specification sheets for configuring building systems, guidelines and training curricula for building trades people and other installers, minimum guidelines for building shell features, or columns on energy applications in building trade publications. Also, it may be useful to facilitate "peer exchange," the recommendation of particular technologies by building industry professionals to their colleagues. This technique has been successfully used by the American Public Power Association to improve conservation programs among public power utilities.

Building product information centers could be set up to supply building managers and contractors with up-to-date, objective information on the performance, cost, reliability, and best applications of common building systems and equipment.

Related Work

Papers by Berberich, Eichenberger, and Frankel from the 1984 ACEEE Santa Cruz conference discuss issues related to technology transfer. Papers and reports from the AIA Foundation (1983), American Society of Heating, Refrigerating, and Air Conditioning Engineers (1985), Science Applications, Inc. (1983), Synergic Resources Corp. (1985), and Brown et al. (1985) are also related. The operation of DOE's Office of Scientific and Technical Information, the Conservation and Renewable Energy Inquiry and Referral Service, the National Appropriate Technology Assistance Service, and the technology transfer activities of EPRI, GRI, and AIA should also provide useful background for additional research.

CONSERVATION PROGRAM EVALUATION

Description of the Research Area

Government and utility conservation programs can be expensive and are often expected to have substantial effects on future energy consumption. Evaluations can help determine whether these programs are worthwhile investments and whether they deliver the "conservation energy resources" expected of them.

The knowledge gained from such evaluations could lead to the wiser expenditure of taxpayer and ratepayer funds and could ensure that these programs closely conform to their original objectives. Currently, decisions about the size, focus, and direction of programs are difficult to make because little relevant empirical evidence exists on which to base such decisions (Fig. 13.4).

Successful integration of program evaluation into the overall program planning, implementation, and analysis cycle would yield other benefits as well. For example, the data collected and behavioral models developed in conducting evaluations would likely prove useful in improving residential and commercial sector energy-demand forecasting models.

Finally, we are aware of no evaluations that compare the relative effectiveness of different types of program interventions. For example, government regulations and utility rebates represent two ways to improve efficiency of new appliances. Unfortunately, little is known about the relative costs, energy savings, market acceptance, and cost-effectiveness of these two approaches.

Issues to Address

A major issue is whether communication between program managers (in utilities, private sector companies, and government agencies) and researchers (analysts and evaluators) can be improved. Such

Fig. 13.4. Existing conservation programs and plans are generally supported by a very narrow information base. Increasing the quantity and quality of evaluations could greatly strengthen this base.

communication would increase the quality and usefulness of evaluations. For example, program managers can identify issues to be clarified in the evaluations, and evaluators can design pilot programs so that their results are statistically valid.

A second issue is how to decide which types of programs require which types of evaluation. A program involving financial incentives may require different evaluative techniques than a mass-media campaign.

A third issue is how to improve the analytical techniques used to assess the energy savings, electric load impacts, and economic effects produced by a program. This question includes such issues as weather normalization of energy use, separation of program and nonprogram energy-use changes, adjustment for self-selection, and transferability of results from one location to another.

A fourth issue is how to measure nonenergy benefits of conservation programs. Claims for conservation programs often include an improvement in environmental quality, an increase in local employment, an improvement in indoor comfort, and so on. Unfortunately, these claims are usually not backed by data and need to be quantified and validated.

A fifth issue is how to integrate outcome and process evaluations. Outcome evaluations determine the effects of programs (e.g., actual energy savings). Process evaluations analyze the mechanics of program operation (e.g., materials used to train appliance salesmen). Integrated evaluations could estimate the contribution of different program components to the overall benefits and costs.

A final issue is how to develop and use appropriate evaluation techniques. That is, how can the staff that is operating a program choose and use evaluation designs that are reliable and useful without assistance from highly trained and experienced program evaluators? Standard evaluation tools would help overcome this problem.

Possible Approaches

"Canned" program evaluation techniques for all major types of government and utility programs could be developed. Utility regulators and research organizations could require that 1 to 5% of all conservation budgets be devoted to evaluation. DOE could sponsor workshops at which program operators and researchers discuss their mutual needs and interests.

Evaluation methods could also be used to assess performance and penetration of new technologies. Results might suggest better ways to hasten commercialization and adoption of these systems.

Additional effort is also needed to ensure that important conservation programs *are* adequately evaluated. Changes in government and industry performance reviews might be considered so that individuals are rewarded for bottom-line results rather than for spending money or for just delivering services. Changes in organizational culture could greatly increase the demand for competent evaluations.

Related Work

The evaluation literature is immense and deals with several of the topics noted above but by and large has not been used to develop evaluation methods for energy conservation programs. Some evaluations of conservation programs are of high quality and can form the basis for other evaluations; projects conducted by the Michigan Energy Administration, Oak Ridge National Laboratory, Pacific Gas and Electric Company, and Southern California Edison can serve as valuable models for others. The proceedings of a recent conference on evaluation (Argonne National Laboratory 1985) contains many valuable papers on evaluation methods and results.

14

LONG-TERM AND VISIONARY RESEARCH
AND PROGRAM PRIORITIES

A. INTRODUCTION

We see changes around us every day. Societal priorities (such as international security, resource uses, economic philosophies, and toxic-waste management) and personal values (such as comfort and lifestyle preferences) shift in response to political institutions, economic conditions, personal well-being, and cultural changes. Technological development introduces new products (such as computers and microwave ovens) that change the ways we think, live, and work. Institutions adjust their roles and the services they offer, as witnessed by the dramatic changes in telecommunications and banking.

As our community and personal worlds change, we can expect the ways we use energy to change also. Some change will involve new functions and some will apply new technologies to current activities. Predicting these changes and anticipating future research needs is difficult.

In this chapter we identify four innovative areas for long-term research. Additional ideas for future work are presented as "one-minute visions."

B. PRIORITY RESEARCH TOPICS

The research topics presented here affect social institutions, building technologies, and human interaction with living and working spaces. Each would require many years to fully implement, and each can benefit from research and testing to assess realistic impacts and feasibility. Some of these research topics are close to a proof of concept test; others require several years of research before a prototype building or program could be tested.

ALTERNATIVE ENERGY PRICING SYSTEMS TO INCREASE ECONOMIC AND ENERGY EFFICIENCY

Description of the Proposed Program

In the past, energy prices were based on average cost, failing to give a clear signal to consumers of the marginal costs incurred by utilities. In the 1970s and 1980s, energy pricing often shifted to inverted rates (where the more energy used, the higher the price paid for each additional unit of energy) and to time-of-use rates. Other rate concepts in use include "conservation rates" for efficient new homes, and line-extension allowances or demand subscription rates for efficient loads. Although these rates reward energy users for efficiency and time-managed use of energy, the pricing system is not tied to energy efficiency investments.

A variety of fuel pricing schemes could be developed and tested for their effects on energy efficiency investments, operation and maintenance practices, and utility revenues. Many such schemes are being considered by utilities throughout the country. Although much activity is underway in this area, fully implementing efficient and equitable energy pricing schemes will take considerable research and experimentation and, therefore, many years to achieve. Here we discuss one such innovative pricing alternative.

For new construction, the designer, builder, or developer exerts tremendous influence over the energy required to operate buildings. Yet subsequent occupants pay for inefficiency. To be effective, the reward and penalty system must be brought forward to the builder. For existing buildings, energy consumers currently billed on an inverted rate schedule or time-of-use rate may not necessarily equate their utility bill with a penalty for energy inefficiency. And customers receiving bills based on average costs have no efficiency reward or penalty system at all.

A new pricing scheme is proposed by which utilities and other energy suppliers could charge their customers variable fuel prices according to the energy efficiency of the customers' buildings. The pricing scheme would apply both to new buildings through a "capital recovery fee" and to existing buildings through a rate schedule that would reward efficiency and penalize inefficiency. The scheme, based on long-run marginal fuel costs, would be designed to dramatically affect builder or customer decisions to invest in energy efficiency.

With new buildings, a builder or developer could be charged a fee that corresponds to the present value of the capital investment the utility would have to make to supply the energy requirements of the building. The fee would assess a substantial cost on inefficient buildings (e.g., $1,000 per kW of demand or $0.80 per annual therm of gas). Presumably, the builder would elect to improve the designed efficiency of the building to reduce the initial fee.

With existing buildings, billing programs could be developed to inform energy users of the relative efficiency of their energy use compared to similar buildings, and a premium charge would be made to utility customers who exceeded a target level of efficiency (Fig. 14.1). The billing program could combine direct monthly feedback to the customer (on both absolute and comparative energy consumption) with a reward or penalty pricing system. Supplemental information could be included with the bill on actions the customer could take to reduce future bills through energy efficiency improvements.

This pricing scheme combines rate concepts and feedback mechanisms that have separately been tested and implemented on a limited scale. The

---

TAPANGA POWER & LIGHT COMPANY
P.O. BOX 100
TAPANGA, KT 01000

June 1989 Electricity Bill For:

CHATWORTH TOWER APARTMENTS
101 Lake Blvd.

Total consumption	125,431 kWh
Base allowance*	96,000
Weather adjustment	8,500
Occupancy adjustment	-3,000
	101,500 @ 5.1¢ = $5176.50
Excess consumption	23,931 @ 7.4¢ = $1770.90
Total Bill	$6947.40

*Typical electricity use for efficient apartment buildings.
Improvements to your heating and water heating systems
could cut electricity bills by about $12,000 a year.

Fig. 14.1. Alternative energy pricing methods might improve efficiency of both new and existing buildings. This hypothetical electricity bill is based on a tariff that charges a higher price for electricity used beyond the building's predetermined efficiency level.

most critical feature is tying information on relative energy use to the monthly bill or capital recovery fee to allow energy decision-makers to evaluate the economic benefits of investing in greater energy efficiency.

Institutional Changes Needed

To achieve energy prices that varied with building energy efficiency, utilities might implement "least-cost energy planning." Rate-design experts would have to incorporate projections of future resource costs (from resource planners and financial analysts) in addition to the traditional focus on current costs of service and revenue requirements. State regulatory agencies would have to adopt long-term planning horizons (e.g., 20 years) in their review of utility investment, services, and rates. And some form of payment mechanism would have to be developed for efficient building practices and community developments that are beneficial to society but not to the energy supplier (e.g., cash payment to a gas utility when a builder designs a passive solar home that requires only backup service that is unprofitable to the utility).

Steps Needed to Establish Proof of Concept
or Feasibility

Different forms of rates and energy efficiency feedback mechanisms that are easy for the builder or customer to understand and that clearly present alternatives for energy efficiency improvement would have to be investigated. The best ways to rate and label buildings to provide energy efficiency information would have to be determined. The frequency with which feedback and reward or penalty charges should be made (e.g., one-time charge for new buildings vs a yearly charge) would need to be assessed. The building-analysis models used to assess energy efficiency would have to be standardized and generally accepted. Improved analytical tools that analyze utility costs and assess the effects of different rate structures on demand, supply, and revenues might be needed. Finally, attention would have to be paid to potential loopholes. For example, a builder might eliminate air-conditioning systems from building plans to achieve "an efficient building," but the buyer or occupant might retrofit the building with inefficient air conditioning at a later date.

Related Work

Stern and Aronson (1984) discuss past efforts to influence energy conservation actions through energy information programs. Conferences sponsored by the California Public Utilities Commission include discussions of revisions to charges for utility service line extension that would incorporate credits for energy efficiency.

LEAST-COST ENERGY PLANNING AND
SERVICE DELIVERY

Description of the Concept

Under least-cost energy planning (LCEP), utilities would not be rewarded on the basis of an allowed rate of return on capital plant investment. Instead, they would be rewarded with higher profits for delivering services that meet customer energy service requirements at the lowest possible cost, thus motivating the utilities to invest in energy efficiency technologies and services when these would be less costly than conventional utility supply resources (Fig. 14.2). Although several utilities and PUCs are discussing LCEP and developing appropriate analytical tools, much more research is needed before such plans are widely implemented throughout the utility industry.

The utility could meet its energy requirements by buying power from whomever it chose (itself, another utility, private power producers, customers with surplus onsite generators, etc.) and by purchasing energy services (heating, lighting, or cooling) from independent ESCOs, from its own ESCO subsidiary, or directly from customers. Customers could choose

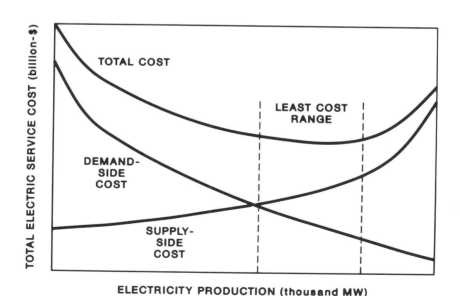

Fig. 14.2. Least-cost energy planning identifies the optimal mix of generation and conservation resources for a given level of energy service.

energy services (e.g., energy efficiency investments, operating and maintenance, or other services) from the provider of their choice. These choices would help customers meet their energy needs at the lowest possible cost. Along with the service options, we could expect to see substantial innovations in energy services and their rate structures.

Institutional Changes Needed

To bring about LCEP and service delivery, state regulations would have to (1) permit utilities to organize into ESCOs, (2) set basic energy-commodity rates, and (3) review the profits of an energy service subsidiary. PUC regulations will require changes to reward utility demand-side investments in a fashion comparable to supply-side investments. Commissions will need to devote considerable effort to the determination of what is meant by least-cost planning. As examples, LCEP could imply minimization of rates (¢/kWh) to customers, minimization of utility revenue requirements, or minimization of customer energy service costs. The relative attractiveness of different options will depend strongly on which optimization criteria are chosen and on the discount rates used to assess future benefits and costs.

The analytical skills of both utility and PUC staff will need to be expanded to adequately assess the full range of supply and demand options available to utilities under LCEP. New planning tools or new ways to integrate current tools may be needed. Utilities may want to reorganize their departments dealing with rates, corporate planning, forecasting, capacity expansion, dispatching, conservation programs, market research, and load research to implement LCEP.

Utility organizations will need to develop a more entrepreneurial outlook because competition with other businesses in the marketplace will require sharp planning, financial analysis, and marketing skills. Potential problems of competition between utilities and local contractors will need to be addressed by state and federal regulators.

Steps Needed to Establish Proof of Concept or Feasibility

Although some utilities have begun to develop and use LCEP models and processes, much more work is needed before utility managers will rely on these analyses for decision making. Additional data, especially on the load-shape impacts of conservation and load management programs, are needed. Efforts are needed to define useful ways to combine individual resource options into portfolios meaningful to utility management. Work is needed to quantify what is meant by "least cost" (to whom and over what time period). Finally, LCEP plans need to be implemented and then evaluated. Such evaluations should identify whether customers obtained improved services at lower cost, whether unfair trade practices were a problem to local businesses, and whether state PUCs were able to adequately review and assess these programs.

Related Work

Several utilities have begun to develop LCEPs, including Pacific Gas and Electric, Sierra Pacific Power, and Northeast Utilities. Their experiences provide valuable insights concerning the data, models, and processes that can help develop meaningful plans. Also, EPRI and others have developed integrated planning models, which incorporate analyses traditionally performed in several utility departments.

PEOPLE-ORIENTED COMFORT SYSTEMS

Description of the Idea

The current approach to providing comfort within buildings is to condition the entire building volume. People-oriented comfort strategies (POCS), however, maintain the building at temperatures and humidity levels based on the needs of building materials and content and provide a separate means of maintaining comfort for occupants. Because current building energy consumption is primarily determined by human needs, such a dual system should consume less energy. The best example of POCS is task lighting, where general illumination is provided through fixtures in the ceiling and lighting for specific tasks is provided with desk lamps at illumination levels controlled by the person. POCS could provide an individual with his or her thermal, luminous, and fresh-air needs, responding to an individual's physiology, activity level, and health.

This idea is not new; localized and/or portable sources of heat, task lamps, hand-held fans, and portable bedwarmers are a few examples of POCs. Today's POCS include the electric blanket, battery-operated garments, task lighting, portable fans, room air conditioners, humidifiers, and desk-top air purifiers. On airplanes, each seat has its own light and fresh-air outlet. For extreme environments, we have astronaut space suits, diver clothing, fire-fighter protective garments, and speciality clothing for hazardous environments.

A people-oriented cooling system was tested in a Kansas factory. The system maintained a satisfactory working environment in the plant even on very hot days. The system also was much less expensive to install and to operate than the alternatives.

Several POCS designs are suggested as possible R&D areas. Portable self-contained units would provide thermal, humidity, lighting, and fresh-air quality levels. Such systems could plug into a supply network within the building, like a telephone. Furniture-integrated systems may be part of a work-station, desk, or chair, with performance controlled by the occupant. Garment-integrated systems could be developed, similar to the special uniforms worn by some workers or athletes.

The nature of building design and building subsystems would change significantly. Construction costs might be lower because of improved net to gross volume ratios made possible by reductions in mechanical equipment and ducts. Because buildings themselves would not provide complete comfort, lighter-weight buildings, less-expensive enclosures, and greater interior flexibility could offer new architectural design opportunities with less material use.

Institutional Changes Needed

Institutions concerned with standards, product safety, health, and human welfare will need to explore their roles vis-a-vis POCS. The professional practices of architects, developers, and engineers may change with the introduction of this new technology.

Steps Needed to Establish Proof of Concept or Feasibility

Prototypes need to be developed to explore the feasibility and energy-saving potential of POCS, user response, and the impact of POCS on support systems and energy-supply networks.

Technology transfer would be critical to the progress of this technology because nonbuilding industries are presently the ones involved in small-scale comfort technologies (e.g., for airplanes, motor homes, and hospitals) and in specialty clothing (e.g., for aerospace, military, and sports).

Health, energy, safety, and comfort standards would have to be reexamined for buildings that use POCS, and comfort standards would need to be developed for the POCS themselves. It is conceivable that building energy consumption standards may be based on occupancy and activity patterns as opposed to generic building type. Several questions would have to be answered: What happens to a person's physiology upon leaving the POCS work space and moving into a building area that is maintained at different temperature levels? What are the appropriate temperatures for the general building volume?

Responsibilities for providing comfort within buildings would need to be redefined. Should the building owner, employer, or employee provide and maintain the POCS?

The impact on building materials and equipment by the maintenance of building volumes at different temperature and humidity levels than now used needs to be investigated.

Finally, research is needed to assess how people react to, and interact with, POCS. Do such systems improve comfort and worker productivity?

Related Work

Institutions such as ASHRAE, IES, NASA, HVAC manfacturers, and aircraft companies are already involved in research related to POCS. Disciplines such as the social sciences, textile design, aerospace medicine, architecture, mechanical engineering, and physiology are critical to any research program related to POCS.

DYNAMIC BUILDINGS AND BUILDING COMPONENTS

Description of the Idea

Current buildings are almost totally static. The few features that respond to the dynamics of climate and human activities are windows, skylights, doors, drapes, venetian blinds, moveable partitions, and some forms of glazing. Shifts in our lifestyles, activity patterns, building occupancy, and building contents suggest an increasing need for highly adaptive buildings.

Dynamic buildings and/or building components could anticipate and respond to changes in nature and human occupancy through changes in the built environment. A high dynamic capability designed into buildings could yield low energy demands, a reduction in the need for material resources, and greater comfort than possible with conventional buildings.

History provides examples of dynamic building elements and structures. Ivy growing on buildings is a dynamic form of shading, keeping the masonry cool in the summer and exposing it to solar radiation in the winter. The Japanese house, with moveable exterior walls and interior partitions, allows one to use space on the cool side of the building in the summer and on the warm side in the winter.

Many dynamic design concepts already exist. They can be organized into dynamic interiors, dynamic building enclosures, and dynamic buildings.

Dynamic interiors permit modification of the spatial volume, its configuration, and degree of enclosure or openness. For example, dynamic elements could allow one to increase the floor-to-ceiling height in the summer and decrease it in the winter or to change the location of interior spaces according to seasons. Ceiling, floor, and wall surfaces could change color, absorption, reflectance, and emissivity in response to changes in luminal and thermal requirements.

Dynamic building enclosures (e.g., roof and wall elements) could change not only their color, reflectance, etc. but also the movement of air through the envelope according to internal needs or exterior conditions. Dynamic devices could automatically provide shading or light reflection or could track solar radiation. The building could modify its exposure and sensitivity to the elements according to weather forecasts. Controls could use microprocessors for automatic adjustment, with manual overrides for greater occupant control.

Dynamic buildings could change their overall form, orientation, configuration, or degree of exposure (Fig. 14.3). A building could have spaces that extend outward from a fixed core that are collapsed when not in use. Alternatively, modules of space could be added to or taken away from a basic building core as required. Old spaces could be traded in and new rooms could be plugged in. The building could raise or lower itself in or out of the ground or water to control its degree of exposure. Floating structures could permit a building to rotate or submerge itself and could also tap the thermal and mechanical energy of the surrounding water.

Institutional Changes Needed

To realize even prototypical examples of dynamic buildings might require a shift in the conceptualization of what buildings mean and how they are used. Building owners, architects, engineers, contractors, and occupants would have to modify established modes of thinking about the building. Building codes and zoning regulations would also need to be modified.

Steps Needed to Establish Proof of Concept or Feasibility

Modeling and simulation tools for dynamic building elements and structures need to be designed and tested. Various enclosure and component prototypes need to be designed and developed along with an assessment of which elements and forces are most critical to control dynamically. Prototype testing, comparative analysis, and validation of modeling tools are also needed. And human responses to dynamic systems need to be understood and tested.

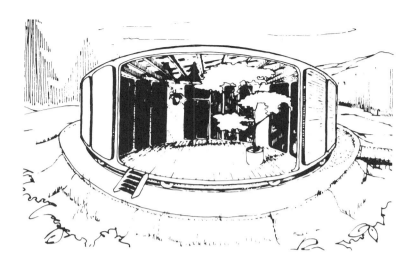

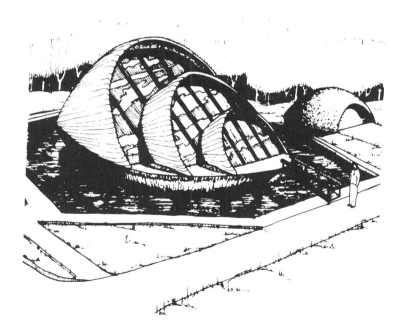

Fig. 14.3. Dynamic buildings could feature ones that revolve on a track or float in the water.

C. VISIONS FOR FUTURE BUILDING
AND COMMUNITY ENERGY USE

As part of the 1984 ACEEE Santa Cruz Conference, we conducted an experiment called "One-Minute Visions of Energy Futures." We invited conference attendees to share their visions of energy futures and to suggest long-term research endeavors. We received about 90 visions, of which seven are presented here.

A NATIONAL TESTING AND DEMONSTRATION
PROGRAM FOR SUPER ENERGY EFFICIENT BUILDINGS

The U.S. data base on extremely efficient buildings is currently limited to results from a handful of programs in a few regions of the country. In addition, few builders or developers have hard evidence that passive solar and energy efficiency techniques work as predicted. This initiative would give grants to developers and builders from all geographic regions of the U.S. to build energy efficient residences alongside their regular stock. A developer building 40 homes on a site, for example, would be paid to make 20 of them super energy efficient and/or passive solar but otherwise similar to the rest. DOE or another organization would monitor the construction cost, the selling price, and the utility bills of all 40 houses. This initiative would provide a comprehensive national data base on the actual costs and benefits of super energy efficient buildings. It would also provide builders and developers hands-on technical information about the construction of such buildings.

LEARNING FROM FAILURES

An annual Energy Awards Contest should be held to identify the most instructive failures. Prizes would be based on three main criteria: how many other people are able to avoid repeating the same mistake by learning from this one, how disastrous a mistake it was, and whether the problem was solved and not merely identified.

Little direct impact would be made on "leading edge" practitioners, who will try new things to make their own mistakes, nor on diehard laggards, who will only respond (slowly and grudgingly) to the whip of energy code requirements. Instead, the willing but slightly cautious "second-round" practitioners, who like to copy innovators, will have a system of warnings about which new building designs and retrofit schemes are failures. By avoiding (or modifying) these schemes, considerable capital, prestige, and confidence in conservation will be conserved for measures that really work.

UNDERSTANDING STILL-EVOLVING PATTERNS
OF ENERGY USE

We are in a time of basic restructuring of the patterns and forces that channel energy through society. Ten years hence, organizations that are barely visible today (e.g., third-party financing firms, ESCOs, and independent power producers) will be major actors. In this time of restructuring, small "fluctuations" will propagate and grow while long-established patterns will decline. In such a transition, standard analytical methods cannot forecast changes accurately. We are not near the equilibrium that is implicitly assumed by analytical methods that use such concepts as "elasticity" and that use averages of large masses of data. More systematic attention is needed to the aberrations, fluctuations, and departures from norms and conventional patterns.

ENERGY SERVICE COMPANIES BECOME AS COMMON
AS FINANCIAL ADVISORS, LAW FIRMS, AND
JANITORIAL SERVICES

We now hire professional brokers and advisors to handle our financial affairs, lawyers to handle our legal affairs, and janitorial services to clean up our messy affairs. Energy costs now account for 11% of total GNP in the U.S. Why don't we have professional companies managing our "energy affairs?" ESCOs are gaining in popularity in the U.S., although they manage energy services for only a very small fraction of the floor space in commercial and residential buildings. In France, on the other hand, "chauffage" or comfort companies sell space heat and hot water for about 70% of the commercial-building floor space.

ESCOs could routinely provide the space heat, hot water, refrigeration, and lighting services for buildings, including small commercial and residential buildings. Consumers would pay for these energy services rather than pay traditional fuel suppliers. Payment would be based on standard methods for monitoring and charging for energy services. Providers of energy services have a natural incentive to invest in conservation to minimize their costs. This incentive would induce third-party ESCOs to replace resistance heating with heat pumps, perform shell retrofits, install efficient lighting systems, and retrofit and maintain HVAC equipment on a massive scale. With energy services metered and charged for, users have an incentive to conserve as well.

THE COMMERCIALIZATION OF "SAVED ENERGY"

Saved energy could become a commodity traded as conventional fuels are traded today. Individual businesses, industries, and homeowners could sell saved energy either to an energy company, a "saved energy" brokerage firm, or another business or consumer. Saved energy could be sold to the highest bidder or sold at some fixed price (like the avoided cost levels now used for cogeneration buybacks). The incentive in all cases would be to avoid building new, costly energy supply facilities. Many utilities would pay for savings to hold down load growth (as is now occurring to a limited extent). One consumer might pay for savings on the property of another to offset his or her own load growth and thereby avoid paying steep marginal-capacity charges for new loads (in a similar manner as air pollution can be offset among emitters under "bubble" provisions). A brokerage firm would trade energy savings for the same reasons.

New York Saved-Energy Exchange		
BUY		
Product	**Amount**	**Offer**
Peak-power savings	40 kW	$200/kW
Natural-gas savings	600 therms/day	$0.50/therm
Base-load		
electricity savings	5000 kWh/mo.	$0.07/kWh
SELL		
Product	**Amount**	**Price**
Fuel-oil saving	150 bbl/day	$28/bbl
Peak-power savings	25 kW/day	$240/kW

NATIONAL CLEAN INDOOR AIR ACT

The existing Clean Air Act protects public health from outdoor air pollution but does not address indoor air quality (IAQ) even though people spend 80 to 90% of their time indoors. Research in recent years has identified a number of air pollutants that can be hazardous in indoor environments.

A federal law could be enacted to protect public health inside buildings. The elements of this act might include:

- Restrictions on the use of particularly hazardous products, such as kerosene heaters, urea-formaldehyde foam insulation, formaldehyde-emitting products, and pesticides;

- Adoption of minimum ventilation requirements for new housing through state and local building codes;

- Authorization for a national indoor radon survey;

- Provision of funds to states for IAQ testing, and for information and education programs; and

- Provision of adequate long-term funding for research on health effects, pollutant identification, and pollutant control techniques.

SUPER ENVELOPES FOR NEW AND EXISTING BUILDINGS

Spatial envelopes are enclosed volumes that temper, modulate, and control the elements of nature for the benefit of their internal spaces. Familiar forms of spatial envelopes include covered arcades, atria, attached sunspaces, galleries, buildings within buildings, courtyards, covered streets, and enclosed malls.

Spatial envelopes can enclose groups of buildings. A large enclosure can protect and modulate natural forces for all the buildings within its enclosure. Such a spatial envelope could collect, store, and control flows of energy, air, and moisture. Such a structure would reduce the design requirements for the individual buildings within the envelope. Interior buildings could be lightweight, and their construction detailing could be simplified.

The large-scale spatial envelope concept could be applied to groups of buildings. In such cases, streets would become malls, and outdoor spaces would become semi-conditioned environments to extend their use in inclement weather. City street and ground maintenance might be reduced, and the buildings themselves could become thermal-storage elements.

15

RESEARCH AND PROGRAM OPPORTUNITIES

A. SHORT-TERM RESEARCH AND PROGRAM NEEDS
WHOLE-BUILDING TOPICS

General

A broad array of whole-building energy performance research could be conducted. That research would include modeling and validation of building energy performance; field performance monitoring and evaluation of whole buildings over extended periods; assessment of building subsystem performance; and investigation of the relative efficiency contribution from individual stages in the building process, such as building design, equipment selection, construction techniques, technology performance, process equipment use, organizational influences, and facility operation and maintenance. In addition, the actual performance of new construction could be investigated to determine if buildings operate as they are designed to.

Whole Buildings

Low-energy commercial buildings could be monitored and data could be compiled about them, including information on how building managers and engineers operate technically sophisticated building equipment and systems.

The importance and cost of improved operation and maintenance of energy using systems could be assessed by measuring the actual performance and degradation of such systems over time; the costs of different levels of routine operation and maintenance; and the frequency of repair for new energy technologies (e.g., heat-pump water heaters and energy-management control systems).

Feedback on building energy performance could be improved by developing techniques to reward building operators for monitoring performance, reporting failures, and correcting deficiencies.

Experiments, field tests, and surveys could be conducted to compare the predicted and actual energy savings of retrofits to existing buildings and to identify the factors that explain these differences.

Research could be conducted to determine:

- Appropriate energy efficient designs for single-family homes in warm climates (both dry and humid),

- The sensitivity of building energy performance to microclimate factors, and

- The impacts of microclimate sensitivity on predictive accuracy.

Field studies could be conducted on the effects of residential appliances and HVAC equipment on building energy performance to determine how design features and usage factors interact, how improved laboratory test procedures relate to performance, and how increasing appliance efficiency affects space conditioning.

Thermal Performance of Building Components

The degradation of energy-conservation features (such as insulation and vapor barriers) could be investigated, and ways of increasing their useful lives could be developed. The thermal-mass characteristics of different building types and components could be analyzed to determine their relationships to comfort, energy efficiency, peak electric loads, and climate. And the impact of building design concepts (such as atria, spatial envelopes, and earth integrated structures) on whole-building energy performance could be analyzed.

Data Compilation and Analysis

Historical trends in building energy performance by building type, fuel, and region could be better understood by documenting the extent of recent changes, their manifestations (technical changes vs operation and maintenance), determinants (fuel prices, economic activity, and conservation programs), and changes in building use or features.

Whole-building energy performance guidelines could be developed for existing buildings on the basis of documented achievable levels of efficiency.

The quality and integration of energy related data collection by DOE/EIA, BLS, Census, Treasury, and other federal agencies could be improved. Major steps in this direction would be:

- Updating the design of EIA's residential and commercial surveys to include questions about respondent attitudes, participation in conservation programs, and recent purchases of energy using equipment, and

- Including two or more years of energy billing data with each survey response.

Coordinated end-use data collection techniques could be developed for use by federal, state, and utility organizations. These techniques could include standard survey questions; standard interviewer protocols (mail, telephone, and on-site); and cooperative conduct of surveys.

Experiments could be conducted to determine the best methods to determine the actual energy savings from retrofitting (e.g., a full year of monthly bills, a few weeks of daily monitoring, or a few hours of intensive monitoring). Techniques for aggregating data from multiple sources (e.g., utility records, building records, surveys, and on-site data collection) could be improved to reduce costs of data collection needed for whole-building energy performance. And methods could be developed to apply technical data and results gained from one location to another situation. Such transfer methods would include adjustments for electric loads, energy costs, climate, building type, and building size.

Techniques could be developed to collect energy data at reasonable costs by:

- Interpreting appliance end-use information from analysis of utility bills or electric line "signature analysis" or

- Employing new technologies, such as fiber optics, remote metering, or microprocessor applications.

Improved data bases could be established within utilities at a sufficient level of precision and reliability for utility and government planning. These data bases could quantify energy use and the impacts of demand-side management programs on the magnitude and timing of use.

Nonenergy Impacts

The advantages and disadvantages of improved energy efficiency in regard to environmental, employment, comfort, sense of well-being, and other factors could be investigated and described.

Research and data collection on the presence of air pollutants in buildings and the implications for health and energy consumption could be expanded to include:

- Monitoring the range of pollutants in commercial buildings,

- Relating those pollutants' presence to HVAC systems and other building parameters,

- Determining the factors affecting pollutant levels in residences,

- Conducting a national radon survey, and

- Investigating the levels of organic compounds in buildings and their health effects.

Techniques could be developed to ensure adequate indoor air quality in buildings with minimal reductions in energy efficiency by eliminating or reducing radon entry into homes; providing controlled ventilation systems for residences both with and without heat recovery; reducing organic-pollutant levels, particularly in mobile homes; standardizing rating procedures for air-to-air heat exchangers; and determining the relationships among ventilation, indoor air quality, and energy use.

Predictive Tools for Professionals and Energy Consumers

Analytical tools that are more powerful and easier to use than currently available techniques could be developed for designing new homes, analyzing HVAC systems (current and new technologies), assessing building energy efficiency improvements, incorporating electricity peak effects into energy-demand models, energy audit programs, and home energy rating systems. Simple calculational tools could be developed to enable building operators to make weather adjustments to utility consumption data for year-to-year or month-to-month comparisons. And a "customized appliance energy prediction tool" could be developed to help consumers assess optimal equipment purchases in relation to likely household use of appliances.

BUILDING TECHNOLOGY TOPICS

Building and Building-Component Performance

Increased research could be conducted on retrofit techniques for multifamily buildings and mobile homes distinct from those developed for single-family buildings. Retrofit techniques for commercial buildings, including HVAC modifications, task lighting, interior partitioning strategies, thermal-zone concepts and shell modifications, could be developed and monitored.

Applied research could be performed on advanced insulation materials and techniques, advanced refrigerant mixtures, innovative windows and window systems, envelope systems (wall and roof assemblies, earth-contact surfaces, and ceilings), and moisture controls.

Techniques could be developed to integrate energy efficiency features into the building structure, to reduce air-conditioning energy use, to ameliorate moisture problems, and to provide passive cooling for homes in hot climates.

Materials

The performance of building materials (including concrete, masonry, wood, gypsum, insulation, vapor retarders, paints and other coating materials, plastics, and glass) could be determined with particular attention to energy-related characteristics, such as thermal capacitance, adsorptivity, emittance, and reflectivity. New materials, such as desiccants and phase-change materials, could be developed and evaluated for both passive and active applications.

The interaction of infiltration, natural and forced ventilation, and joining techniques could be analyzed.

Equipment

Motors and low-cost motor controllers could be improved. Waste-heat recovery systems could be developed for furnaces, refrigerators, air conditioners, and boilers. Conventional and new HVAC components could be integrated with passive energy systems, and advanced controls and sensors for both active and passive systems could be developed.

Small-scale cogeneration and other decentralized, integrated energy efficiency technologies (thermal energy storage, fuel cells, etc.) could be improved and their application expanded, especially if their size and weight could be reduced.

Community-scale energy technologies that might substitute for individual building equipment could be identified, and the minimum densities of buildings needed to justify them could be determined.

Computer-controlled building management systems could be devised to save energy and reduce peak demand. Control technologies (direct or computerized) that utilities could rely on for load management and low-cost metering equipment for time-of-use or real-time energy pricing could be developed.

Customer-programmable and -controllable technology could be produced to allow customers to choose time-of-use rates, interruptible rates, contract curtailment, or other energy pricing alternatives. Such technologies might include programmable appliance controls, demand subscription service, thermal storage, group-load curtailment controls, microprocessing and communications technologies, "smart homes," or "smart meters." In-building energy meters that are programmable by individual appliance usage on a time-of-day basis could be developed. Energy rates could be set on a fixed schedule or on a real-time basis, where instantaneous costs are charged according to the utility's energy-source mix (i.e., marginal cost).

Dehumidifiers and high efficiency HVAC systems that include heat recovery and that can be used at various scales (workstation, room, zone, building) could be developed.

More efficient residential appliances (refrigerators, freezers, air conditioners, fuel-fired water heaters, fuel-fired heat pumps, and lighting) could be developed along with integrated microprocessor controls for their operation.

BEHAVIORAL TOPICS

Planning Tools for Demand-Side Energy Management

Recent experiences with top-down institution-oriented planning for energy efficiency (e.g., Northwest Power Plan and DOE state programs) and with bottom-up planning based on energy-user needs (e.g., Minneapolis Neighborhood Energy Watch, Santa Monica Energy Fitness Program, and Massachusetts Energy Federation efforts) could be assessed to find out what these experiences tell us about program planning, program implementation, and user response.

Large-scale experiments that involve cooperation among diverse groups (utilities, state regulators, environmentalists) could be studied and assessed. The Hood River Conservation Project in Oregon would provide a useful starting point for such an assessment.

The ability of states, regions, or localities to develop and implement least-cost energy plans could be studied, and the impact that such plans are having on overall energy use and supply decisions could be assessed.

Analytical tools could be developed to compare the use of demand-side management options with that of supply-side resources in utility resource planning. Such tools could incorporate market dynamics (customer responses to price, new technologies, comfort, etc.) and produce market penetration estimates, load forecasts, and utility economic and financial projections.

Models of energy management decision-making could be developed for each market segment (e.g., existing residences, owner-occupied homes, tenant-occupied homes, seniors on limited incomes, and growing families) to guide policy decisions and strategy formulation.

The traditional models for forecasting energy end uses cannot account for the effects of new technologies and programs. The changes needed in these models to allow such accounting could be investigated. In addition, residential and commercial energy demand models could be adapted to incorporate effects of consumer attitudes and decision making not based on engineering or economic factors. Such models could be used to test the validity of implicit discount rates for energy investment decisions and to integrate results from program evaluations.

Institutional and Organizational Influences on Energy Use Decisions

The incentives for and barriers to greater attention to energy efficiency among various institutions could be systematically identified. These institutions include state energy offices, public utility commissions, electric and gas utilities, financiers, architects, engineers, state and local building code officials, equipment and building-component manufacturers, contractors, appliance dealers, and product retailers.

Organizations or institutions that influence energy efficiency could be identified by market segment, and how they might sponsor various "interventions" (information, incentives, policies, or regulations) could be assessed.

Mechanisms could be developed to improve access to capital and/or to lower capital and transaction costs for energy efficiency investment, including:

- Vendor financing,
- Bank/insurance pools,
- State or federal funds in direct-loan pools,
- State or federal funds for leveraging,
- Interest subsidies, and
- Grants.

Barriers to Energy Efficiency

Behavioral and decision-process research could be conducted to identify market imperfections that hinder:

- Increased investment in residential energy-efficient technologies by market segments,

- Greater adoption of energy-efficient "practices" in residences,

- Greater investment in energy-efficient commercial building technologies, and

- Greater adoption of energy-efficient operating and maintenance practices in commercial buildings.

The degree of "institutionalization" of energy management could be assessed by identifying and monitoring the existence of a recognized profession of energy managers or by determining the degree to which energy management becomes integrated into a repertoire of building management skills.

Ways of motivating more cost-effective purchase behavior related to residential energy consumption could be researched with a study of consumer behavior and how consumers react to information and incentives about energy efficiency.

The perceived risks of energy management actions could be identified, and techniques could be developed to overcome them for:

- Individual consumers (information on peer experience and guarantees),

- Institutional energy users (considering performance contracts and salary guarantees), and

- Investors (considering short payback measures).

Such research could investigate the roles and effectiveness of different methods to reduce risk to building owners, the implementation costs of different approaches, and the participation rates and energy savings for different groups.

Strategies to Overcome Barriers

Government programs and research funding could be adopted to place greater emphasis on results than on program operation. This shift might involve changes in performance review systems to encourage program managers to focus on "bottom line" results rather than on activities.

The effects and costs of marketing strategies for residential conservation programs could be tested with different strategies offered to different market segments. And models could be established for governments or utilities to determine when energy services should be stimulated and when market demand is sufficient to sustain self-supporting commercial services.

Regulatory Issues

Compliance to building codes and other regulatory programs needs to be evaluated, and ways to improve compliance could be developed. The impacts of standards on equipment manufacturers, building designers, utilities, low-income energy users, and others could be further studied.

Alternative forms of building and appliance standards (e.g., performance vs prescriptive or fleet average vs minimum standards) could be investigated in terms of energy efficiency, ease and cost of compliance and enforcement, structural changes within industry, and capital cost.

Better regulatory tools could be developed to stimulate and reward "least-cost resource plans" by utilities. These tools could establish reasonable planning and analysis time frames, encourage entrepreneurial roles for utilities and their subsidiaries, reward utility management and shareholders for socially desirable actions, and ensure that cost-effectiveness tests are appropriate to societal objectives.

INTERVENTION STRATEGY TOPICS

Research on Information Transfer

A nationwide energy conservation clearinghouse could be established to improve transfer of information (both technical and programmatic) from researchers to practitioners and from one implementing organization to another.

Forums for researchers and practitioners to exchange information and experiences could be created to increase the likelihood that research recognizes program operator needs, that operators have the opportunity to apply research findings promptly, and that researchers are aware of one another's work. The major targets for energy efficiency information (e.g., lenders, contractors, architects) could be identified, and their explicit needs for and ways of using information could be determined.

The results of performance testing of energy devices and appliances could be disseminated to utilities, program operators, and individual consumers. And a central information source could be established to communicate detailed program operator experiences to other program operators.

Support of Program Delivery

How can government agencies and utilities better stimulate marketing of conservation equipment and services? To answer this question, the relevant knowledge of consumer behavior from other fields (e.g. consumer product research, psychology, and anthropology) could be compiled for use by energy program designers and operators. That compilation could identify target markets, types of information or intervention strategy called for, and applications of communication and consumer decision theory.

A guide to the attributes of successful conservaton programs could be compiled with step-by-step guidelines on how to plan and execute such programs; "canned" descriptions of model programs could be prepared. The economics of retrofits (e.g., the costs for do-it-yourself vs contractor installation) could be examined. And the benefits and costs of improved training for building inspectors, auditors, contractors, and others could be investigated.

Support of Energy Services Delivery

A "client focus" on the part of program operators could define the market and identify appropriate marketing techniques. "Trade allies" for government agencies and utilities could be identified among builders, manufacturers, financial institutions, consumer organizations, and environmental organizations. Ways to reduce overhead costs of programs could be identified. And tools could be developed to help commercial energy users evaluate alternative financing mechanisms (e.g., internal financing, shared savings, and leasing).

Research on Incentives

Answers could be sought for the following questions: How can incentives be used to alter investment payback levels, grab consumer attention, accelerate conservation action, and guide consumers in choosing the most efficient equipment or products? How can the incremental impacts of incentive programs be evaluated best? What are the effects of incentives in terms of participation rates, level of investments made, actual energy savings, program implementation costs, and program cost-effectiveness? And can utility "capacity factor" and connection charges act as economic incentives for the design and construction of efficient new buildings?

Evaluation of Program Strategies and Interventions

Evaluation of government and utility energy-conservation programs could be improved by:

- Integrating evaluation into program planning and implementation,

- Improving communication between evaluators and program managers,

- Conducting evaluations that focus on program process (how the program is delivered) as well as program effects in an integrated fashion, and

- Developing methods to determine the evaluability of programs.

Standard analytical methods are needed so that program operators who otherwise would not perform quality evaluations can readily evaluate program designs and results. Available aids should provide or explain survey instruments, use of focus groups, selection of experimental groups, interpretation of energy bill information, and other common techniques.

Model evaluation methods should outline the generic types of evaluation, test acceptance of program design, test causality of program activities on outcomes, determine relation of outcomes to program objectives, and describe how to avoid common problems (e.g. self-selection, reliance on self-reports, and creation of control groups).

Guidelines and methods could be developed to distinguish among the effects of price elasticity, technology improvements, and conservation programs and among accelerated, new, and naturally occurring conservation efforts.

Cooperative planning and evaluating efforts among energy researchers and program operators could be encouraged and supported to improve the quality of energy programs; an important step in their improvement could be the establishment of standardized means for reporting program experiences.

Selected Experimentation with Intervention Strategies

Existing federal and state programs could experiment with intervention strategies more frequently. Specifically, they could investigate program delivery issues, information exchange and networking among program operators, business planning for program operators to increase overall quality and success of programs, and leverage of limited funds.

Ways to improve energy savings relative to the level of effort could be investigated with small-scale studies. Options to consider include improving the energy performance of conservation measures taken (a technical solution), persuading energy users to take more actions and increasing program participation rates (marketing solutions), and reducing overhead costs (management solutions).

The relative cost-effectiveness of various levels of conservation treatment could be assessed. For example, the relative cost-effectiveness of low (<\$300), medium (\$800 to 1500), or high (>\$2,000) conservation investments could be investigated, and the optimal levels of investment by climate, building type, and other relevant factors could be determined. Ways to improve the effectiveness of residential audit and information programs could be investigated. The impact of energy ratings and labels for buildings and appliances could be evaluated as could the impact of voluntary guidelines for architects and builders.

Alternative policy options for ensuring adequate indoor air quality could be developed and evaluated. Such options might include regulatory programs (e.g. bans on unvented space heaters and limits on the use of formaldehyde products in mobile homes); the use of ventilation guidelines and standards; and labeling programs.

B. LONG-TERM RESEARCH AND PROGRAM NEEDS

GOVERNMENT POLICIES

Federal and state governments could establish ramped performance standards for buildings on a long-range basis (e.g., for 1990, 1995, and 2000). Funding would be needed for near-term training and professional exchanges on energy efficient design, building materials, and construction practices. Research would also be required to develop the building materials and analytical techniques needed to meet the future standards.

Urban policies need to be reviewed and updated with respect to solar-access laws, wind-access laws, access to daylight, restrictions on thermal dumping and pollution from buildings, and credits for recycling instead of demolishing buildings.

Government policies and programs that affect energy supply and demand need to be carefully assessed for appropriate balance. These assessments should consider the economics of various demand and supply alternatives from the perspectives of both energy users and producers. In addition, nonenergy impacts (such as environmental quality, comfort, and individual control) should also be considered and quantified.

Governments could tax inefficient appliances, with the tax set equal to the value of the energy wasted (including the social investments) over the lifetime of the equipment. And state and federal governments could establish institutional planning frameworks, regulatory methods, and/or incentives to stimulate purchase of efficient appliances.

UTILITY ACTIVITIES

The legal and regulatory issues related to utility operation of energy service companies could be researched. Utilities could experiment with owning and installing decentralized technologies on customer premises (e.g., cogeneration, fuel cells, thermal-energy storage, photovoltaic cells, and high-efficiency appliances).

Utilities could investigate the use of differential energy pricing or "value of service" rate concepts, where customers would choose the desired reliability of service (e.g., amount available or interruptibility). Under such a system, a higher price would be charged for high-reliability service. Energy prices set according to the relative efficiency of building energy use could also be investigated, with efficient buildings getting lower rates. The rates could be designed to incorporate investments the utility must make to provide energy services to the building.

Deregulation that would permit distribution utilities to buy energy resources from a variety of sources could be investigated. These sources could include small power producers, ESCOs, and cogenerators as well as conventional energy sources. The role of unregulated energy service subsidiaries within regulated electric and gas utilities could also be investigated. Such subsidiaries offer the potential of improved services and lower costs to energy consumers because of the likely competition among private and utility ESCOs. However, the potential for abuse by large utilities should be carefully assessed.

ENERGY SERVICE BUSINESS OPPORTUNITIES

Turnkey building-operation services could be tested to see if energy efficiency can be improved when specific responsibility is given to an outside entity on a profit basis. Candidates for experiments include government facilities and nonprofit institutions. Information about the relative efficiency of individual buildings could be made available to identify good prospects for services, and the information vendors could charge for their services.

HUMAN BEHAVIOR AND LIFESTYLES

New forms of building occupancy mix and building use patterns could be investigated to improve energy efficiency. For example, instead of continuing the separation of residential and commercial buildings, zoning could integrate particular types of commercial buildings with housing. A building that is primarily internal-load dominated and occupied during the day might be merged with a residential building so the load can be shed in the evening.

The impacts on energy use of new concepts in work and living arrangements could be analyzed. Such concepts include working at home, sharing work spaces through staggered work schedules, and integrating cottage industries with housing.

Additional research on the fundamentals of individual and organizational decision-making is needed. Such research could examine both operating decisions and capital investment decisions. Particular attention could be paid to the likely effectiveness of different government interventions to improve the energy efficiency of both investment and operating decisions.

TECHNOLOGIES FOR APPLICATION TO WHOLE COMMUNITIES

Sets of buildings could have their energy systems linked so they could share and transfer advantages derived from their individual orientations or internal-load characteristics. Such shared energy sources might include active, passive, and photovoltaic solar systems and wind-energy conversion systems.

Super envelopes (large enclosures of groups of buildings) could be developed and tested to assess the impacts of various devices for solar collection, natural ventilation, and thermal storage.

Floating communities and buildings could be tested to see if they can economically use renewable forms of energy and/or use thermal gradients within waterbodies.

NEW TYPES OF BUILDINGS

A demonstration program of super energy efficient buildings could be conducted to provide data and to increase professional knowledge about energy efficient designs.

Earth integrated commercial and multifamily buildings could be investigated in various urban contexts and climate zones.

Prototype buildings with skin-dominated forms could be constructed to determine the potential energy efficiency, productivity, environmental quality and/or comfort improvement. Such prototypes might provide greater access to natural heating, cooling, and daylighting, decreasing the need for nonrenewable energy.

"Smart walls" capable of adjusting themselves to optimize comfort and energy consumption and dynamic buildings able to track renewable forms of energy and avoid extreme forces of nature could be developed and tested.

TECHNOLOGIES FOR EXISTING BUILDINGS

The energy benefits of building recycling could be analyzed. Such recycling might include reuse of the building in the original form and place, reuse of the building in a different location, reuse of the building materials at the original site, or reuse of the building materials at another site.

Community-based cogeneration systems with potential for waste recycling, long-term energy storage, or combustible-waste recycling could be developed and tested.

Biologically controlled envelope energy-flow controls could be developed, and their impact on building energy consumption could be assessed. The energy conservation potential of HVAC systems that use waste heat could be investigated for installation in individual buildings or building zones. And integrated dynamic energy-flow controls could be developed and tested to modulate light, air, sound, and thermal flows for building envelopes.

New materials to improve building energy performance could be investigated. The potential energy and nonenergy impacts of using smaller-scale furnishings, equipment, and foldaway furnishings to reduce overall building volume could be researched to see if this would improve energy efficiency. And task-oriented comfort systems could be developed to moderate energy use and enhance comfort. These technologies could include workstation components, furnishings, mobile systems, and clothing.

A tool that predicts or compares energy use could be created to challenge people to improve the energy efficiency of their homes or commercial buildings. The tool could substitute for or supplement conventional building energy audits. With such a tool, building, appliance, and household characteristics would be plugged into a computational model, the model would process this information, and it would offer a personalized plan of action to save energy and money.

AFTERWORD

We began this book with a question. Does it make sense to publish a book on energy conservation at a time when the energy "crisis" has receded from public view? When even some energy professionals consider the nation's energy problems over? Our answer was and still is a rousing yes!

In our view, the 10 to 15 years after the 1973 oil embargo showed the tremendous contribution that improved energy efficiency can make to the resolution of our nation's energy problems. U.S. energy use today is roughly one-third lower than it would have been had pre-embargo trends continued. Improved energy efficiency can be considered the largest source of "new" energy production since 1973, having provided more energy services than new oil and gas wells, new coal mines, or new electric power plants. Indeed, greater energy efficiency throughout the world is a principal reason why we are now enjoying stable energy prices.

Further, these efficiency improvements were not just "one shot" quick fixes. Although a tremendous amount has been accomplished since 1973 (as documented in Part II), much more remains to be done (as discussed in Part III). Technologies, systems, institutions, and conservation programs continue to evolve and advance rapidly.

Consider lighting as an example. During the past few years, several manufacturers have developed compact fluorescent light bulbs for use in residential lighting fixtures, replacing the usual incandescent bulbs. A typical fluorescent lamp uses only 18 W, but delivers the light output of a conventional 75-W incandescent bulb, a whopping 75% decline in electricity requirements. As production and competition among manufacturers increase, the performance of these bulbs will improve and their costs will surely decline. Additional efforts are needed to ensure that these high-efficiency bulbs are purchased and used.

Advances in lighting are also important in commercial buildings, in part because the heat generated by lights contributes to summer air-conditioning loads and to electric utility system peaks. Use of high efficiency fluorescent lamps, high frequency ballasts, improved fixtures, and control systems that make the best use of available daylight can cut lighting electricity use by up to 60%. As these systems are improved and as experience is gained with their design and operation, electricity savings will also increase.

The typical government or utility energy conservation program of the 1970s focused on owners of single-family homes. By today's standards, these programs, with their emphasis on generalized lists of ways to save energy, were weak. Today, many utilities operate programs that pay much of the cost of their customers' efficiency improvements (generally retrofits to existing homes and the purchase of high efficiency appliances). Some utilities now offer substantial energy conservation and load management programs for their commercial customers, including free energy audits and financial assistance for the installation of measures recommended during the audits. Utilities and governments have become much more skillful in identifying appropriate technologies, in marketing their programs to particular subpopulations, and in measuring the actual performance of these programs. This progress in program operation has made it easier to justify conservation program budgets.

Some state public utility commissions and the utilities they govern are exploring regulatory changes that focus on least-cost energy planning. The purpose of these deliberations and studies is to adjust the "rules of the game" to ensure socially desirable energy futures that are also profitable to utilities. In addition, some utilities have formed subsidiaries to offer energy services in an effort to reduce overall energy costs, to compete with other energy service companies and fuel suppliers, and to increase profits. These regulatory and institutional changes, as well as other emerging steps, suggest important future changes in the functions and regulation of U.S. utilities.

Experience during the past decade shows that the patterns and determinants of energy use in buildings are very complicated. The relationships among energy efficient technologies and their performance characteristics, purchase and construction decisions, operation and maintenance practices, and individual and organizational behavior are as important as they are difficult to understand. Knowledge about these issues has increased substantially during the past several years, in both the conduct of research projects and in the operation of conservation programs.

Because of these complexities and interactions among technologies and behavior, the importance of increased collaboration among professionals with different academic backgrounds is now well recognized. Such collaboration should involve cooperation among utilities, government agencies, community groups, manufacturers, and professional societies. It also requires closer integration of research methods and implementation techniques.

We end this book with a restatement of our two major themes. First, the U.S. has made terrific strides in improving building energy efficiency during the past 10 to 15 years. Researchers and practitioners have learned a great deal about the technologies and programs appropriate for encouraging greater efficiency gains. Second, the future is as bright as the past. We are on the threshold of implementing recent breakthroughs in energy-efficient building designs and technologies on a vast scale. Additional cost-effective ways to improve energy efficiency are waiting to be discovered, developed, and implemented. As Pogo said, "We are faced with insurmountable opportunities."

REFERENCES

CHAPTER 1

Adams, R. C., et al. August 1984. *A Retrospective Analysis of Energy Use and Conservation Trends: 1972–1982,* PNL-5026, Pacific Northwest Laboratory. Richland, Wash.

Dickey, D. F., et al. 1984. "Effects of Utility Incentive Programs for Appliances on the Energy Efficiency of Newly Purchased Appliances," *Doing Better: Setting an Agenda for the Second Decade,* American Council for an Energy-Efficient Economy, Washington, D.C.

Department of Energy May 1985. *Energy Use Trends in the United States, 1972–1984,* draft, Office of Policy Integration, Washington, D.C.

Energy Information Administration 1984a. *Monthly Energy Review,* DOE/EIA-0035, Energy Information Administration, U.S. Department of Energy, Washington, D.C.

Energy Information Administration October 1984b. *Energy Conservation Indicators 1983 Annual Report,* DOE/EIA-0441(83), Energy Information Administration, U.S. Department of Energy, Washington, D.C.

Energy Information Administration April 1985. *Annual Energy Review 1984,* DOE/EIA-0384(84), Energy Information Administration, U.S. Department of Energy, Washington, D.C.

Geller, H. S. 1985. "Progress in the Energy Efficiency of Residential Appliances and Space Conditioning Equipment," in *Energy Sources: Conservation and Renewables,* American Institute of Physics, New York.

Goldstein, D. B. 1984. "Efficient Refrigerators in Japan: A Comparative Survey of American and Japanese Trends Towards Energy Conserving Refrigerators," *Doing Better: Setting an Agenda for the Second Decade*, American Council for an Energy-Efficient Economy, Washington, D.C.

Klepper, M. 1984. "Issues and Challenges for Third Party Financing: An Agenda for the Next Ten Years," *Doing Better: Setting an Agenda for the Second Decade*, American Council for an Energy-Efficient Economy, Washington, D.C.

Hirst, E., et al. 1983. "Recent Changes in U.S. Energy Consumption: What Happened and Why," *Annual Review of Energy, 8,* Annual Reviews, Palo Alto, Calif.

Pacific Northwest Laboratory October 1984. *Residential and Commercial Buildings Data Book,* DOE/RL/01830-16, Richland, Wash.

CHAPTER 2

City of Austin August 1984. *Austin's Conservation Power Plant,* Resource Management and Electric Utility Departments, Austin, Tex.

Cooper, M. N., and Sullivan, T. L. 1983. *Equity and Energy: Rising Energy Prices and the Living Standards of Lower Income Americans,* Westview Press, Boulder, Colo.

Energy Information Administration November 1984. *Residential Energy Consumption Survey, Consumption and Expenditures, April 1982 Through March 1983,* DOE/EIA-0321/1(82), Energy Information Administration, U.S. Department of Energy, Washington, D.C.

Energy Information Administration April 1985. *Annual Energy Review 1984,* DOE/EIA-0384(84), Energy Information Administration, U.S. Department of Energy, Washington, D.C.

Electrical World September 1984. "35th Annual Electric Utility Industry Forecast," *Electrical World* **198** (9), 49–56.

Garey, R. B., and Stevenson, W. March 1983. "Estimated Employment Effects of the Department of Energy's Weatherization Assistance Program," Testimony presented to the Subcommittee on Energy Conservation and Power, U.S. House of Representatives.

Geller, H. S. August 1985. *Energy Demand and Conservation in the U.S.: 1970–1984,* American Council for an Energy-Efficient Economy, Washington, D.C.

Goldenberg, J., et al. 1985. "An End-Use Oriented Global Energy Strategy," *Annual Review of Energy, 10,* Annual Reviews, Palo Alto, Calif.

Kelly, H. November-December 1983. "Energy Conservation and National Security," *Energy Conservation Bulletin* 3(3), 1,4–6.

National Audubon Society 1984. *The Audubon Energy Plan 1984,* National Audubon Society, New York.

National Real Estate Investor May 1984. "Energy Costs a Problem for Building Owners," *National Real Estate Investor.*

Northwest Power Planning Council April 1983. *Northwest Conservation and Electric Power Plan,* Northwest Power Planning Council, Portland, Oreg.

Sebold, F. D., Thayer, M. A., and Hageman, R. K. 1983. *Assessment of the External and Intangible Effects of SDG&E's 1983 Conservation and Load Management Programs,* Regional Economic Research, San Diego, Calif.

Solar Energy Research Institute 1981. *A New Prosperity: Building a Sustainable Energy Future,* Brick House Publishers, Andover, Mass.

Spiewak, I., et al. December 1983. *Energy Conservation R&D Priority Analysis,* ORNL/PPA-83/9, Oak Ridge National Laboratory, Oak Ridge, Tenn.

Williams, R. H., et al. 1983. "Future Energy Savings in U.S. Housing," *Annual Review of Energy, 8,* Annual Reviews, Palo Alto, Calif.

Williams, R. H. February 1985. *A Low Energy Future for the U.S.,* PU/CEES-186, Center for Energy and Environmental Studies, Princeton University, Princeton, N.J.

CHAPTER 3

American Institute of Architects 1982. *Energy in Design: Techniques,* A Level 2 Workshop of the AIA Energy Professional Development Program, Washington, D.C.

American Institute of Architects 1981. *Energy in Design: Application,* A Level 3 Workshop of the AIA Energy Professional Development Program, Washington, D.C.

ASHRAE 1977. *ASHRAE Handbook and Product Directory, 1977 Fundamentals,* American Society of Heating, Refrigerating, and Air-Conditioning Engineers, Inc., New York.

Givoni, B. 1976. *Man, Climate, and Architecture,* Applied Science, London.

Givoni, B. 1986. *Passive Cooling for Buildings,* McGraw Hill, New York.

Givoni, B., and Milne, M. 1981. *The AIA Energy in Design Technique,* Washington, D.C.

Office of Technology Assessment March 1982. *Energy Efficiency of Buildings in Cities,* USGPO, Washington, D.C.

Olgyay, V. 1963. *Design with Climate,* Princeton University Press, Princeton, N.J.

Watson, D. 1979. *Energy Conservation Through Building Design,* McGraw Hill, New York.

CHAPTER 4

American Institute of Architects Research Corporation January 1978. *Phase One/Base Data for the Development of Energy Performance Standards for New Buildings, Residential Data Collection and Analysis,* U.S. Department of Energy, Washington, D.C.

Blumstein, C., et al. April 1980. "Overcoming Social and Institutional Barriers to Energy Conservation," *Energy* 5(4), 355–372.

Bureau of the Census December 1982, *Statistical Abstract of the United States, 1982-83,* U.S. Department of Commerce, Washington, D.C.

Busch, J. F., et al. 1984. "Measured Heating Performance of New, Low-Energy Homes: Updated Results from the BECA-A Database," *Doing Better: Setting an Agenda for the Second Decade,* American Council for an Energy-Efficient Economy, Washington, D.C.

Chandra, S., Fairey, P., and Houston, M. September 1983. "A Handbook for Designing Ventilated Buildings," *Principles of Low Energy Building Design in Warm, Humid Climates,* FSEC-GP-21-83, Florida Solar Energy Center, Cape Canaveral, Fla.

Clark, E., and Burdahl, P. 1980. "Radiative Cooling: Resource and Applications," presented at the Fifth National Passive Solar Conference, Amherst, Mass.

Cleary, P. 1984. "Humidity in Attics—Source and Control Methods," *Doing Better: Setting an Agenda for the Second Decade,* American Council for an Energy-Efficient Economy, Washington, D.C.

Corum, K. R. 1984. "Interaction of Appliance Efficiency and Space Conditioning Loads: Application to Residential Energy Demand Projections," *Doing Better: Setting an Agenda for the Second Decade,* American Council for an Energy-Efficient Economy, Washington, D.C.

Duke Power Co. October 1983. *Review of Federal Policies and Building Standards Affecting Energy Conservation in Housing,* Serial No. 98–85, pp. 634–643, USGPO, Washington, D.C.

Energy Information Administration October 1984. *Energy Conservation Indicators 1983 Annual Report,* DOE/EIA-0441(83), U.S. Department of Energy, Washington, D.C.

Fairey, P. W. December 1982. "Effects of Infrared Radiation Barriers on the Effective Thermal Resistance of Building Envelopes," presented at the ASHRAE/DOE Conference on Thermal Performance of Exterior Envelopes of Buildings II, Las Vegas, Nev.

Fairey, P. W. September 1983. "Building Design Considerations," in *Principles of Low Energy Building Design in Warm, Humid Climates,* FSEC-GP-21-83, Florida Solar Energy Center, Cape Canaveral, Fla.

Givoni, B. November 1981. "Cooling of Buildings by Passive Systems," presented at the First International Passive and Hybrid Cooling Conference, Miami, Fla.

Harris, J., and Blumstein, C., eds. 1984. *What Works: Documenting Energy Conservation in Buildings,* American Council for an Energy-Efficient Economy, Washington, D.C.

Holt, D. August 1984. "Superinsulated Houses: The Importance of Resale Value," *Doing Better: Setting an Agenda for the Second Decade,* American Council for an Energy-Efficient Economy, Washington, D.C.

Housing Assistance Council March 21, 1984. Testimony before the Subcommittee on Housing and Community Development, Hearing to Review Federal Policies and Building Standards Affecting Energy Conservation in Housing, U.S. House of Representatives, pp. 272–288, USGPO, Washington, D.C.

Hutchins, P. F., and Hirst, E. October 1978. *Engineering-Economic Analysis of Mobile Home Thermal Performance,* ORNL/CON-28, Oak Ridge National Laboratory, Oak Ridge, Tenn.

Hutchinson, M., Nelson, G., and Fagerson, M. 1984. "Measured Thermal Performance and the Cost of Conservation for a Group of Energy-Efficient Minnesota Homes," *What Works: Documenting Energy Conservation in Buildings,* ed. by J. Harris and C. Blumstein, American Council for an Energy-Efficient Economy, Washington, D.C.

Laquatra, J. August 1984. "Valuation of Household Investment in Energy-Efficient Design," *Doing Better: Setting an Agenda for the Second Decade,* American Council for an Energy-Efficient Economy, Washington, D.C.

Lischkoff, J. K., and Lstiburek, J. W. 1984. "Building Science Practice and the Airtight Envelope," *Doing Better: Setting an Agenda for the Second Decade*, American Council for an Energy-Efficient Economy, Washington, D.C.

Meyers, S., and Schipper, L. May 1984. "Energy in American Homes: Changes and Prospects," *Families and Energy: Coping with Uncertainty*, Conference Proceedings, ed. by B. M. Morrison and W. Kempton, Michigan State University, East Lansing, Mich.

Mills, E. August 1984. "Raising the Energy Efficiency of Manufactured Housing," *Doing Better: Setting an Agenda for the Second Decade*, American Council for an Energy-Efficient Economy, Washington, D.C.

Mineral Insulation Manufacturers Association March 21, 1984. Testimony before the Subcommittee on Housing and Community Development, Hearing to Review Federal Policies and Building Standards Affecting Energy Conservation in Housing, U.S. House of Representatives, pp. 644–646, USGPO, Washington, D.C.

O'Neal, D. L., Corum, K. R., and Jones, J. L. April 1981. *An Estimate of Consumer Discount Rates Implicit in Single-Family Housing Construction Practices*, ORNL/CON-62, Oak Ridge National Laboratory, Oak Ridge, Tenn.

Ribot, J. C., et al. 1984. "Monitored Low-Energy Houses in North America and Europe: A Compilation and Economic Analysis," *What Works: Documenting Energy Conservation in Buildings*, ed. by J. Harris and C. Blumstein, American Council for an Energy-Efficient Economy, Washington, D.C.

Rouse, R. E. 1983. *Passive Solar Design for Multifamily Buildings*, Massachusetts Executive Office of Energy Resources, Boston, Mass.

Schipper, L., Meyers, S., and Kelly, H. 1985. *Coming in from the Cold: Energy-Wise Housing in Sweden*, Seven Locks Press, Cabin John, Md.

Wendt, R. L. April 1982. *Earth-Sheltered Housing, An Evaluation of Energy-Conservation Potential*, ORNL/CON-86, Oak Ridge National Laboratory, Oak Ridge, Tenn.

Williams, R. H., Dutt, G. S., and Geller, H. S. 1983. "Future Energy Savings in U.S. Housing," *Annual Review of Energy, 8*, Annual Reviews, Palo Alto, Calif.

CHAPTER 5

Adams, R. C., et al. February 1984. *A Retrospective Analysis of Energy Use and Conservation Trends: 1972–1982*, draft, PNL-5026, Pacific Northwest Laboratory, Richland, Wash.

Bleviss, D. L., and Gravitz, A. A. October 1984. *Energy Conservation and Existing Rental Housing*, Energy Conservation Coalition, Washington, D.C.

Bureau of the Census December 1982. *Statistical Abstract of the United States, 1982–83*, 103rd edition, U.S. Department of Commerce, Washington, D.C.

Cleary, P. August 1984. "Humidity in Attics—Sources and Control Methods," *Doing Better: Setting an Agenda for the Second Decade*, American Council for an Energy-Efficient Economy, Washington, D.C.

Crane, L. T. January 1984. *Residential Energy Conservation: How Far Have We Progressed and How Much Farther Can We Go?*, prepared by the Congressional Research Service for the Subcommittee on Energy Conservation and Power, U.S. House of Representatives, USGPO, Washington, D.C.

Crenshaw, R. and Clark, R. E. September 1982. *Optimal Weatherization of Low-Income Housing in the U.S.: A Research Demonstration Project*, NBS Building Science Series 144, National Bureau of Standards, Washington, D.C.

DaSilva, J. P., and Waintroob, D. S. August 1984. "The Mass-Save Multi-Family Experience: Lessons Learned over the Past Three Years," *Doing Better: Setting an Agenda for the Second Decade*, American Council for an Energy-Efficient Economy, Washington, D.C.

Dutt, G., et al., June 1982. *The Modular Retrofit Experiment: Exploring the House Doctor Concept*, PU/CEES No. 130, Princeton University, Center for Energy and Environmental Studies, Princeton, N.J.

Dutt, G. S., Jacobson, D., and Socolow, R. H. February 1983. *Air Leakage Reduction and Energy Savings in the Modular Retrofit Experiment*, PU/CEES No. 113, Princeton University, Center for Energy and Environmental Studies, Princeton, N.J.

Egel, K. S. 1984. "Energy Cost Control in Nonprofit Multifamily Housing: Evaluation of an Energy Extension Service Community Program," *What Works: Documenting Energy Conservation in Buildings*, ed. by J. Harris and C. Blumstein, American Council for an Energy-Efficient Economy, Washington, D.C.

Energy Information Administration August 1984. *Residential Energy Consumption Survey: Housing Characteristics 1982,* DOE/EIA-0314(82) U.S. Department of Energy, Washington, D.C.

Fels, M. F. and Goldberg, M. L. May 1984. "Using Billing and Weather Data to Separate Thermostat from Insulation Effects," *Energy* 9(5), 439–446.

Gathers, W. E., Jr. 1984. "The Alliance to Save Energy/Institute for Human Development Low-Income Heating System Retrofit Program," *Doing Better: Setting an Agenda for the Second Decade,* American Council for an Energy-Efficient Economy, Washington, D.C.

Gold, C. S. 1984. "The Page Homes Demonstration Energy Conservation Computer System," *What Works: Documenting Energy Conservation in Buildings,* ed. by J. Harris and C. Blumstein, American Council for an Energy-Efficient Economy, Washington, D.C.

Goldman, C. A. 1984. "Measured Energy Savings from Residential Retrofits: Updated Results from the BECA-B Project," *Doing Better: Setting an Agenda for the Second Decade,* American Council for an Energy-Efficient Economy, Washington, D.C.

Hirst, E., and Goeltz, R. March 1984. *Comparison of Actual and Predicted Energy Savings in Minnesota Gas-Heated Single-Family Homes,* ORNL/CON-147, Oak Ridge National Laboratory, Oak Ridge, Tenn.

Hirst, E., White, D., and Goeltz, R. 1985. "Indoor Temperature Changes in Retrofit Homes," *Energy* 10(7), 861–870.

Hirst, E., White, D., and Goeltz, R. November 1983. *Comparison of Actual Electricity Savings with Audit Predictions in the BPA Residential Weatherization Pilot Program,* ORNL/CON-142, Oak Ridge National Laboratory, Oak Ridge, Tenn.

Holt, D. 1984. "Superinsulated Houses: The Importance of Resale Value," *Doing Better: Setting an Agenda for the Second Decade,* American Council for an Energy-Efficient Economy, Washington, D.C.

Johnson, R. C. July 1981. *Housing Market Capitalization of Energy-Saving Durable Good Investments,* ORNL/CON-74, Oak Ridge National Laboratory, Oak Ridge, Tenn.

Kempton, W., and Montgomery, L. October 1982. "Folk Quantification of Energy," *Energy* 7(10), 817–827.

Kempton, W., et al. 1984. "Do Consumers Know 'What Works' in Energy Conservation?" *What Works: Documenting Energy Conservation in Buildings,* ed. by J. Harris and C. Blumstein, American Council for an Energy-Efficient Economy, Washington, D.C.

Kensill, F. 1984. "Measuring Savings in Retrofitted Oil Heating Equipment," *Doing Better: Setting an Agenda for the Second Decade,* American Council for an Energy-Efficient Economy, Washington, D.C

Meier, A., Wright, J., and Rosefeld, A. H. 1983. *Supplying Energy Through Greater Efficiency, The Potential for Conservation in California's Residential Sector,* University of California Press, Berkeley, Calif.

Meyers, S., and Schipper, L. May 1984. "Energy in American Homes: Changes and Prospects," *Families and Energy: Coping with Uncertainty,* ed. by B. M. Morrison and W. Kempton, Michigan State University, East Lansing, Mich.

Mills, E. 1984. "Raising the Energy Efficiency of Manufactured Housing," *Doing Better: Setting an Agenda for the Second Decade,* American Council for an Energy-Efficient Economy, Washington, D.C.

Office of Technology Assessment March 1982, *Energy Efficiency of Buildings in Cities,* USGPO, Washington, D.C.

O'Neal, D. L., Corum, K. R., and Jones, J. L. April 1981. *An Estimate of Consumer Discount Rates Implicit in Single-Family Housing Construction Practices,* ORNL/CON-62, Oak Ridge National Laboratory, Oak Ridge, Tenn.

Proctor, J. 1984. "Low Cost Furnace Efficiency Improvements," *Doing Better: Setting an Agenda for the Second Decade,* American Council for an Energy-Efficient Economy, Washington, D.C.

Solar Energy Research Institute 1981. *A New Prosperity: Building a Sustainable Energy Future,* Brick House Publishing, Andover, Mass.

Stern, P. C., and Aronson, E., eds. 1984. *Energy Use, The Human Dimension,* W. H. Freeman and Company, New York, N.Y.

Wagner, B. S. 1984. "Verification of Buildings Energy Use Models: A Compilation and Review," *Doing Better: Setting an Agenda for the Second Decade,* American Council for an Energy-Efficient Economy, Washington, D.C.

CHAPTER 6

A. D. Little, Inc. March 1982. *Consumer Products Efficiency Standards Engineering Analysis Document,* DOE/CE-0030, U.S. Department of Energy, Washington, D.C.

Association of Home Appliance Manufacturers July 1984. *1983 Energy Consumption and Efficiency Data for Refrigerators, Refrigerator-Freezers and Freezers,* Association of Home Appliance Manufacturers, Chicago.

California Energy Commission July 1983. *California's Appliance Standards: An Historical Review, Analysis, and Recommendations*, P400-83-020, California Energy Commission, Sacramento, Calif.

Chang, Y-M. L., and Grot, R. A. July 1979. "Field Performance of Residential Regrigerators and Combination Refrigerator-Freezers," NBSIR 79-1781, National Bureau of Standards, Washington, D.C.

Clear, R. D., and Goldstein, D. B. May 1980. *A Model for Water Heater Energy Consumption and Hot Water Use: Analysis of Survey and Test Data on Residential Hot Water Heating*, LBL-10797, Lawrence Berkeley Laboratory, Berkeley, Calif.

Corum, K. R. 1984. "Interaction of Appliance Efficiency and Space Conditioning Loads: Application to Residential Energy Demand Projections," *Doing Better: Setting an Agenda for the Second Decade*, American Council for an Energy-Efficient Economy, Washington, D.C.

DeCicco, J. M., et al. August 1984. "Energy Scorekeeping for a Multi-Family Building: a Study of Lumley Homes," *Doing Better: Setting an Agenda for the Second Decade*, American Council for an Energy-Efficient Economy, Washington, D.C.

Department of Energy July 1983. *Supplement to: March 1982 Consumer Products Efficiency Standards Engineering Analysis and Economic Analysis Documents*, DOE/CE-0045, U.S. Department of Energy, Washington, D.C.

Dobyns, J. E., and Blatt, M. H. May 1984. *Heat Pump Water Heaters*, EPRI EM-3582, Electric Power Research Institute, Palo Alto, Calif.

Duro-Test Corp. April 1982. *Energy-Efficient Incandescent Lamp: Final Report*, LBL-14546, Lawrence Berkely Laboratory, Berkeley, Calif.

Fairchild, P. D. August 1985. "ORNL Electric Systems Program Overview," *Proceedings of the DOE/ORNL Heat Pump Conference*, CONF-841231, U.S. Department of Energy, Oak Ridge, Tenn.

Fairey, P. April 1984. "Superinsulation and Cooling," presented at the 1984 Superinsulation Forum, Rochester, Minn.

Fechter, J. V., et al. March 1979. *Kitchen Range Energy Consumption*, NBSIR 78-1556, National Bureau of Standards, Washington, D.C.

Florida Solar Energy Center September 1983. *Principles of Low Energy Building Design in Warm, Humid Climates*, Florida Solar Energy Center, Cape Canaveral, Fla.

Gas Research Institute October 1984. *Technology Profile—Advanced Four-Burner Cooktop*, Gas Research Institute, Chicago.

Geller, H. S. June 1983. *Energy Efficient Appliances*, American Council for an Energy-Efficient Economy and the Energy Conservation Coalition, Washington, D.C.

Geller, H. S. August 1984. "Efficient Residential Appliances and Space Conditioning Equipment: Current Savings Potential, Cost Effectiveness and Research Needs," *Doing Better: Setting an Agenda for the Second Decade*, American Council for an Energy-Efficient Economy, Washington, D.C.

Geller, H. S. 1985. "Progress in the Energy Efficiency of Residential Appliances and Space Conditioning Equipment," *Energy Sources: Conservation and Renewables,* ed. by D. Hafemeister, H. Kelly, and B. Levi, American Institute of Physics, New York.

Geller, H. S., et al. February 1986. *Residential Conservation Power Plant Study: Phase I—Technical Potential,* draft, American Council for an Energy-Efficient Economy, Washington, D.C.

Goldstein, D. B. August 1984. "Efficient Refrigerators in Japan: A Comparative Survey of American and Japanese Trends Towards Energy Conserving Refrigerators," *Doing Better: Setting an Agenda for the Second Decade*, American Council for an Energy-Efficient Economy, Washington, D.C.

Good Housekeeping Institute October 1982. *Home Appliance Study*, Good Housekeeping Institute, New York.

Government Printing Office 1981. *Code of Federal Regulations*, Title 10, Part 430, USGPO, Washington, D.C.

Harris, J., and Blumstein, C., eds. 1984. *What Works: Documenting Energy Conservation in Buildings*, American Council for an Energy-Efficient Economy, Washington, D.C.

Hewett, M. J., and Peterson, G. August 1984. "Measured Energy Savings from Outdoor Resets in Modern, Hydronically Heated Apartment Buildings," *Doing Better: Setting an Agenda for the Second Decade*, American Council for an Energy-Efficient Economy, Washington, D.C.

Khattar, M. K. April 1985. *Analysis of Air-reheat Systems and Application of Heat Pipes for Increased Dehumidification,* FSEC-PF-76-85, Florida Solar Energy Center, Cape Canaveral, Fla.

Kweller, E., and Silberstein, S. June 1984. *Performance and Selection Criteria for Mechanical Energy Saving Retrofit Options for Single-Family Residences*, NBSIR 84-2870, National Bureau of Standards, Washington, D.C.

Lannus, A. August 1985. "EPRI Heat Pump R&D Overview," *Proceedings of the DOE/ORNL Heat Pump Conference*, CONF-841231, U.S. Department of Energy, Oak Ridge, Tenn.

Lawrence, W. T. August 1982. *Field Test Measurements of Energy Savings from High Efficiency, Residential Electric Appliances*, Florida Public Service Commission, Tallahassee, Fla.

Lennox June 1984. Press release and product information on the HS14 Series Power Saver Air Conditioners, Lennox Industries, Dallas.

Levins, W. P. April 1980. *Energy and the Laundry Process,* ORNL/CON-41, Oak Ridge National Laboratory, Oak Ridge, Tenn.

Levins, W. P. January 1982. *Estimated Seasonal Performance of a Heat Pump Water Heater Including Effects of Climate and In-House Location,* ORNL/CON-81, Oak Ridge National Laboratory, Oak Ridge, Tenn.

Levins, W. P. August 1984. "An Assessment of the Energy Saving Potential of Nonazeotropic Refrigerant Mixtures," *Doing Better: Setting an Agenda for the Second Decade,* American Council for an Energy-Efficient Economy, Washington, D.C.

Linteris, G. T. 1984. "Performance of Retrofitted and New High Efficiency Gas Equipment: Some Recent GRI Projects," *What Works: Documenting Energy Conservation in Buildings,* ed. by J. Harris and C. Blumstein, American Council for an Energy-Efficient Economy, Washington, D.C.

Ludvigson, V., and VanValkenburg, T. 1978. "Microwave Energy Consumption Tests," *Proceedings of the Conference on Major Home Appliance Technology for Energy Conservation,* CONF-780238, U.S. Department of Energy, Washington, D.C.

Meier, A., and Whittier, J. August 1984. "Purchasing Patterns of Energy-Efficient Refrigerators and Implied Consumer Discount Rates," *What Works: Documenting Energy Conservation in Buildings,* ed. by J. Harris and C. Blumstein, American Council for an Energy-Efficient Economy, Washington, D.C.

Meier, A., et al. 1983. *Supplying Energy Through Greater Efficiency—The Potential for Conservation in California's Residential Sector,* University of California Press, Berkeley, Calif.

Middleton, M. G., and Sauber, R. S. September 1983. *Research and Development of Energy Efficient Appliance Motor-Compressors, Vol. IV—Production Demonstration and Field Test,* ORNL/Sub/78-7229/4, report prepared by Kelvinator Compressor Co. for Oak Ridge National Laboratory, Oak Ridge, Tenn.

Norgard, J. S., and Heeboll J. November 1983. "Radical Reduction in Domestic Need for Electricity, Exemplified by the Refrigerator," presented at the ECC Seminar on Energy Saving in Buildings, The Hague.

Palmiter, L., and Kennedy, M. September 1983. "Annual Thermal Utility of Internal Gains," presented at the 8th National Passive Solar Conference, American Solar Energy Society, Santa Fe, N.M.

Reid, F. A. April 1981. *Characteristics, Efficiencies, and Use of Appliances in Residential Market Segments,* Lawrence Berkeley Laboratory, Berkeley, Calif.

Ruderman, H., Levine, M. D., and McMahon, J. E. August 1984. "Energy Efficiency Choice in the Purchase of Residential Appliances," *Doing Better: Setting an Agenda for the Second Decade,* American Council for an Energy-Efficient Economy, Washington, D.C.

Science Applications Inc. March 1982. *Consumer Products Efficiency Standards Economic Analysis Document,* DOE/CE-0029, U.S. Department of Energy, Washington, D.C.

Stern, P. C., and Aronson, E., eds. 1984. *Energy Use—The Human Dimension,* W. H. Freeman and Co., New York.

Thrasher, W. H., et al. August 1984. "Development of a Space Heater and a Residential Water Heater Based on the Pulse Combustion Principle," *Doing Better: Setting an Agenda for the Second Decade,* American Council for an Energy-Efficient Economy, Washington, D.C.

Topping, R. F. December 1982. *Development of a High-Efficiency, Automatic-Defrosting Refrigerator-Freezer, Phase II-Field Test,* Vol. III, ORNL/SUB/77-7255/3 Oak Ridge National Laboratory, Oak Ridge, Tenn.

Usibelli, A. August 1984. "Monitored Energy Use of Residential Water Heaters," *Doing Better: Setting an Agenda for the Second Decade,* American Council for an Energy-Efficient Economy, Washington, D.C.

Verderber, R. R., and Rubinstein, F. R. June 1983. *Comparison of Technologies for New Energy-Efficient Lamps,* LBL-16225, Lawrence Berkeley Laboratory, Berkeley, Calif.

CHAPTER 7

American Institute of Architects 1982. *Energy in Design: Techniques,* A Level 2 Workshop of the AIA Energy Professional Development Program, Washington, D.C.

Andersson, B., et al. 1984. "Energy Effects of Electric Lighting Control Alternatives in Response to Daylighting," *Doing Better: Setting an Agenda for the Second Decade,* American Council for an Energy-Efficient Economy, Washington, D.C.

Department of Energy 1983. *Passive Solar Commercial Building Program, Case Studies,* DOE/CE-0042, U.S. Department of Energy, Washington, D.C.

Energy Information Administration March 1983a. *Nonresidential Buildings Energy Consumption Survey: 1979 Consumption and Expenditures Part 1*, DOE/EIA-0318/1, Energy Information Administration, U.S. Department of Energy, Washington, D.C.

Energy Information Administration December 1983b. *Nonresidential Buildings Energy Consumption Survey: 1979 Consumption and Expenditures Part 2*, DOE/EIA-0318(79)/2, Energy Information Administration, U.S. Department of Energy, Washington, D.C.

Gardiner, B. L., et al. August 1984. "Measured Results of Energy Conservation Retrofits in Non-Residential Buildings: An Update of the BECA-CR Data Base," *Doing Better: Setting an Agenda for the Second Decade*, American Council for an Energy-Efficient Economy, Washington, D.C.

Gordon, H. T., Estoque, J., Hart, G. K., and Kantrowitz, M. March 1985. "Performance Overview: Passive Solar Energy for Non-Residential Buildings," Solar Energy Group, Lawrence Berkeley Laboratory, Berkeley, Calif.

Harris, J. August 1984. Lawrence Berkeley Laboratory, Berkeley, Calif., personal communication to Walter Kroner.

Harris, J., and Blumstein, C. eds. 1984. *What Works: Documenting Energy Conservation in Buildings*, American Council for an Energy-Efficient Economy, Washington, D.C.

Hirst, E., Carney, J., and Knight, P. June 1981. *Energy Use at Institutional Buildings: Disaggregated Data and Data Management Issues*, ORNL/CON-73, Oak Ridge National Laboratory, Oak Ridge, Tenn.

Int-Hout, D. August 1984. "Environmental Control with Variable Air Volume Systems," *What Works: Documenting Energy Conservation in Buildings*, ed. by J. Harris and C. Blumstein, American Council for an Energy-Efficient Economy, Washington, D.C.

Katrakis, J., and Becker, D. August 1984. "Energy Savings in Buildings of Neighborhood-Based Non-Profit Organizations," *Doing Better: Setting an Agenda for the Second Decade*, American Council for an Energy-Efficient Economy, Washington, D.C.

Kowalczyk, D. April 1985. "Commercial Sector Energy Efficiency Improvements: An Economic and Statistical Analysis of Empirical Data," *Energy Policy* 13(2), 169–178.

MacDonald, M., Goldenberg, D., and Hudgins, E. June 1982. *RD&D Opportunities for Large Air Conditioning and Heat Pump Systems*, ORNL/Sub/80-13817/1&20, Oak Ridge National Laboratory, Oak Ridge, Tenn.

Novey, G. T. August 1984. "Engineering Conservation Opportunities Through the Application of Direct Digital Control of Heating, Ventilation and Air Conditioning," *Doing Better: Setting an Agenda for the Second Decade*, American Council for an Energy-Efficient Economy, Washington, D.C.

New York State Energy Research and Development Authority May 1984. *A Business Guide to Energy Performance Contracting*, Report 84-11, New York State Energy Research and Development Authority, Albany, N.Y.

Office of Technology Assessment March 1982. *Energy Efficiency of Buildings in Cities*, USGPO, Washington, D.C.

Parken, W. H., et al. June 1982. *Strategies for Energy Conservation in Small Office Buildings*, NBSIR 82-2489, National Bureau of Standards, Washington, D.C.

Progressive Architecture April 1982a. "A Baseline for Energy Design," *Progressive Architecture* 63(4).

Progressive Architecture June 1982b. "Energy Design of Office Buildings," *Progressive Architecture* 63(6), 109–113.

Progressive Architecture February 1983a. "Energy to Recover," *Progressive Architecture* 64(2), 136–139.

Progressive Architecture March 1983b. "Energy, A Class by Itself," *Progressive Architecture* 64(3).

Schultz, D. K. August 1984. "End-Use Consumption Patterns and Energy Conservation Savings in Commercial Buildings," *Doing Better: Setting an Agenda for the Second Decade*, American Council for an Energy-Efficient Economy, Washington, D.C.

Selkowitz, S., Arasteh, D., and Johnson, R. August 1984. "Peak Demand Savings from Daylighting in Commercial Buildings," *Doing Better: Setting an Agenda for the Second Decade*, American Council for an Energy-Efficient Economy, Washington, D.C.

Stoops, J., et al. August 1984. "A Baseline for Energy Design," *What Works: Documenting Energy Conservation in Buildings*, ed. by J. Harris and C. Blumstein, American Council for an Energy-Efficient Economy, Washington, D.C.

Sullivan D. August 1984. "A Decade of Energy Management at the Ohio State University," *Doing Better: Setting an Agenda for the Second Decade*, American Council for an Energy-Efficient Economy, Washington, D.C.

Usibelli, A., et al. February 1985. "Commercial Sector Conservation Technologies," LBL-18543, Lawrence Berkeley Laboratory, Berkeley, Calif.

Verderber, R. R., and Rubinstein, F. M. 1984. "Mutual Impacts of Lighting Controls and Daylighting Applications," *Energy and Buildings* **6,** 133–140.

Wall, L., and Flaherty, J. 1984. "A Summary Review of Building Energy Use Compilation and Analysis (BECA) Part C: Conservation Progress in Retrofitted Commercial Buildings," *What Works: Documenting Energy Conservation in Buildings,* ed. by J. Harris and C. Blumstein, American Council for an Energy-Efficient Economy, Washington, D.C.

Wall, L. W., et al. August 1984. "A Summary Report of BECA-CN: Buildings Energy Analysis of Energy Efficient New Commercial Buildings," *Doing Better: Setting an Agenda for the Second Decade,* American Council for an Energy-Efficient Economy, Washington, D.C.

Warren, M. L. August 1984. "Energy Conservation Through Effective Building Control: Closing the Feedback Loop," *Doing Better: Setting an Agenda for the Second Decade,* American Council for an Energy-Efficient Economy, Washington, D.C.

CHAPTER 8

Archer, D., et al. 1984. "Energy Conservation and Public Policy: The Mediation of Individual Behavior," *Doing Better: Setting an Agenda for the Second Decade,* American Council for an Energy-Efficient Economy, Washington, D.C.

Breed, D. P., and Michaelson, M. L. 1984. "Guideline for the Public Sector: Implementing a Third-Party Financed Energy Services Transaction," *Doing Better: Setting an Agenda for the Second Decade,* American Council for an Energy-Efficient Economy, Washington, D.C.

Condelli, L., et al. June 1984. "Improving Utility Conservation Programs: Outcomes, Interventions, and Evaluations," *Energy* **9**(6), 485–494.

Cramer, J. C., et al. May 1984. "The Determinants of Residential Energy Use: A Physical-Social Causal Model of Summer Electricity Use," *Families and Energy: Coping with Uncertainty,* Michigan State University, East Lansing, Mich.

Dubin, J. A., May 1982. *Economic Theory and Estimation of the Demand for Consumer Durable Goods and Their Utilization: Appliance Choice and the Demand for Electricity,* MIT-EL 82-035WP, MIT Energy Laboratory, Cambridge, Mass.

Energy Conservation Bulletin May-June 1985. "Minnesota First with Successful Shared Savings Program for Single Family Homes," Energy Conservation Coalition, Washington, D.C.

Fagerson, M. H., 1984. "Analysis of Lifestyle Factors in Heating Use of New and Weatherized Minnesota Homes," *Doing Better: Setting an Agenda for the Second Decade*, American Council for an Energy-Efficient Economy, Washington, D.C.

Fay, B. 1984. "Voluntary Rental Living Unit Program," *Doing Better: Setting an Agenda for the Second Decade*, American Council for an Energy-Efficient Economy, Washington, D.C.

Feldman, S. 1984. "Why Is It So Hard to Sell 'Savings' as a Reason for Energy Conservation?" *Doing Better: Setting an Agenda for the Second Decade*, American Council for an Energy-Efficient Economy, Washington, D.C.

Fels, M. F., and Kempton, W. May 1984. *Toward More Informative Energy Bills,* PU/CEES 164, Princeton University, Center for Energy and Environmental Studies, Princeton, N.J.

Goett, A. A. May 1983. *Appliance System and Fuel Choice: An Empirical Analysis of Household Investment Decisions*, Electric Power Research Institute, Palo Alto, Calif.

Hausman, J. A. 1979. "Individual Discount Rates and the Purchase and Utilization of Energy-Using Durables," *The Bell Journal of Economics* **10**(1), 33–54.

Hirst, E., and Goeltz, R. 1985. "Accuracy of Household Self-Reports: Energy Conservation Surveys," *Social Science Journal* **22**(1), 19–30.

Hobbs, C. D., et al. 1984. "Energy Management in the Commercial Sector: The Marketing Role of Financing," *Doing Better: Setting an Agenda for the Second Decade*, American Council for an Energy-Efficient Economy, Washington, D.C.

Kempton, W. 1984. "Residential Hot Water: A Behaviorally-Driven System," *Doing Better: Setting an Agenda for the Second Decade*, American Council for an Energy-Efficient Economy, Washington, D.C.

Kempton, W., and Krabacher, S. 1984. "Thermostat Management: Intensive Interviewing Used to Interpret Instrumentation Data," *Doing Better: Setting an Agenda for the Second Decade*, American Council for an Energy-Efficient Economy, Washington, D.C.

Kempton, W., et al. 1984. "Do Consumers Know What Works in Energy Conservation?" *What Works: Documenting Energy Conservation in Buildings*, ed. by J. Harris and C. Blumstein, American Council for an Energy-Efficient Economy, Washington, D.C.

Klepper, M. 1984. "Issues and Challenges for Third Party Financing: an Agenda for the Next Ten Years," *Doing Better: Setting an Agenda for the Second Decade*, American Council for an Energy-Efficient Economy, Washington, D.C.

Kline, J. 1982. "Seattle Bankers, Appraisers Join in Innovative Residential Rating Program," *Energy and Housing Report* II(12), 3.

Kuliasha, M. A., et al. 1985. *Field Performance of Residential Thermal Storage Systems,* EM-4041, Electric Power Research Institute, Palo Alto, Calif.

New York State Energy Research and Development Authority January 1984. *Performance Contracting for Energy Efficiency: An Introduction With Case Studies*, Technical Development Corporation for the New York State Research and Development Authority, Albany, N.Y.

Owens-Corning July 1984. "Do Mortgage Lenders Favor Energy-Efficient Homes?" Owens-Corning Fiberglas, Toledo, Ohio.

Pilati, D. A. February 1975. *The Energy Conservation Potential of Winter Thermostat Reductions and Night Setback*, ORNL-NSP-80, Oak Ridge National Laboratory, Oak Ridge, Tenn.

Ruderman, H., et al. 1984. "Energy Efficiency Choice in the Purchase of Residential Appliances," *Doing Better: Setting an Agenda for the Second Decade*, American Council for an Energy-Efficient Economy, Washington, D.C.

Science Applications, Inc. March 1982. *Consumer Products Efficiency Standards Economic Analysis Document*, DOE/CE-0029, U.S. Department of Energy, Washington, D.C.

Schuck, L., and Millhone, J. August 1984. "Defining Energy Efficiency in Residential Lending Practices," *Doing Better: Setting an Agenda for the Second Decade*, American Council for an Energy-Efficient Economy, Washington, D.C.

Shinn, R. A., and Rametta, A. J. 1984. "A Penny Saved Is Half a Penny Earned: Pennsylvania's Third-Party Financing Experience for Energy Conservation," *Doing Better: Setting an Agenda for the Second Decade*, American Council for an Energy-Efficient Economy, Washington, D.C.

Sonderregger, R. C. 1978. "Movers and Stayers: The Resident's Contribution to Variation Across Houses in Energy Consumption for Space Heating," *Saving Energy in the Home, Princeton's Experiments at Twin Rivers*, ed. by R. H. Socolow, Ballinger Publishing Co., Cambridge, Mass.

Sorensen, J. H. April 1985. "Information and Energy Conserving Behavior," Chap. 3 in *Past Efforts and Future Directions for Evaluating State Energy Conservation Programs,* ORNL-6113, Oak Ridge National Laboratory, Oak Ridge, Tenn.

Sterling Hobe October 1984. *Comparative Analysis of U.S. and Selected Foreign Household Appliance Industries,* U.S. Department of Energy, Washington, D.C.

Stern, P. C. 1984. *Improving Energy Demand Analysis,* National Research Council, National Academy Press.

Stern, P. C., and Aronson, E., 1984. *Energy Use: The Human Dimension,* National Research Council, W. H. Freeman and Company, New York.

Train, K. E. March 1985. *Discount Rates in Consumers' Energy-Related Decisions: A Review of the Literature,* Cambridge Systematics, Inc., Cambridge, Mass.

Vine, E. L., et al. November 1982. "The Applicability of Energy Models to Occupied Houses: Summer Electric Use in Davis," *Energy* 7(11), 909–926.

Weedall, M. 1984. "The Emerging Role of the Public Sector in Third-Party Finance," *Doing Better: Setting an Agenda for the Second Decade,* American Council for an Energy-Efficient Economy, Washington, D.C.

Weihl, J. S. 1984. "Family Schedules and Residential Energy Consumption Behaviors," *Doing Better: Setting an Agenda for the Second Decade,* American Council for an Energy-Efficient Economy, Washington, D.C.

Weisenmiller, R. B. 1984. "A Perspective of the History of Third-Party Financing of Conservation in the United States," *Doing Better: Setting an Agenda for the Second Decade,* American Council for an Energy-Efficient Economy, Washington, D.C.

Wilk, R. R., and Wilhite, H. L. 1984. "Why Don't People Weatherize Their Homes?: An Ethnographic Explanation," *Doing Better: Setting an Agenda for the Second Decade,* American Council for an Energy-Efficient Economy, Washington, D.C.

CHAPTER 9

Association of Home Appliance Manufacturers June 1984. "Comments by AHAM to the Department of Energy on the Proposed Rulemaking Regarding Energy Efficiency Standards," Association of Home Appliance Manufacturers, Chicago.

Anderson, C. D., and Claxton, J. D. 1981. "Energy Labels for Appliances," *Consumers and Energy Conservation: International Perspectives on Research and Policy Options*, ed. by J. D. Claxton, D. C. Anderson, J. R. Ritchie, and G. H. G. McDougall, Praeger, New York.

Anderson, C. D., and Claxton, J. D. September 1982. "Barriers to Consumer Choice of Energy Efficient Products," *Journal of Consumer Research* **9**, 163–170.

Bakken, W. September 1984. California Energy Commission, School and Hospital Program manager, personal communication to Jeanne Clinton.

Baumgardner, E., and Schultz, D. November 1981. *Local Energy Planning Handbook*, P-400-81-036, California Energy Commission, Sacramento, Calif.

Bleviss, D. L., and Gravitz, A. A., 1984. *Energy Conservation and Existing Rental Housing*, Energy Conservation Coalition, Washington, D.C.

Brennan, M. A., and Zelinski, R. W. August 1984. "Performance Contracting in New York City-Owned Multi-Family Buildings," *Doing Better: Setting an Agenda for the Second Decade*, American Council for an Energy-Efficient Economy, Washington, D.C.

Brummitt, M. J. D. August 1984. "Marketing Conservation Through Grassroots Organizing: Neighborhood Energy Workshop Program in Minneapolis," *Doing Better: Setting an Agenda for the Second Decade*, American Council for an Energy-Efficient Economy, Washington, D.C.

California Energy Commission May 24, 1983a. *Rationale for Staff Proposal on Nonresidential Building Standards*, P400-83-017, California Energy Commission, Sacramento, Calif.

California Energy Commission July 1983b. *California's Appliance Standards: An Historical Review, Analysis and Recommendations*, P400-83-020, California Energy Commission, Sacramento, Calif.

California Energy Commission December 1983c. *California's Solar, Wind and Conservation Tax Credits*, P103-83-001, California Energy Commission, Sacramento, Calif.

Capehart, G. L., et al. October 1982. *Florida's Electric Future: Building Plentiful Supplies on Conservation*, Florida Conservation Foundation, Winter Park, Fla.

City of Austin, August 1984. *Austin's Conservation Power Plant*, Resource Management and Electric Utility Departments, Austin, Tex.

Collins, N. E., et al. August 1985. *Past Efforts and Future Directions for Evaluating State Energy Conservation Programs*, ORNL-6113, Oak Ridge National Laboratory, Oak Ridge, Tenn.

Congressional Record September 13, 1984. U.S. House of Representatives, E3816.

Cornwall, B. October 1984. California Energy Extension Service, personal communication to Jeanne Clinton.

Cowell, S. L., and Rebitzer, R. August 1984. "Beyond Technology: Energy Conservation Delivery Systems that Work," *Doing Better: Setting an Agenda for the Second Decade*, American Council for an Energy-Efficient Economy, Washington, D.C.

Crenshaw, R., and Clark, R. E. September 1982. *Optimal Weatherization of Low-Income Housing in the U.S.: A Research Demonstration Project*, NBS Building Science Series 144, National Bureau of Standards, Washington, D.C.

Department of Energy November 1979. *Notice of Proposed Rulemaking, Energy Performance Standards for New Buildings*, DOE/CS-0112, Office of Conservation and Solar Energy, U.S. Department of Energy, Washington, D.C.

Department of Energy April 1983. *An Evaluation of the Institutional Conservation Program—Results of On-site Analysis*, U.S. Department of Energy, Washington, D.C.

Department of Energy January 1984a. *Residential Conservation Service Evaluation Report*, DOE/CE-0086, U.S. Department of Energy, Washington, D.C.

Department of Energy February 24, 1984b. *Residential Energy Conservation Tax Credit Initiatives,* draft report, U.S. Department of Energy, Washington, D.C.

Department of Energy January 1984c. *Energy Conservation Technology R&D Program Plan*, Office of Conservation and Renewable Energy, U.S. Department of Energy, Washington, D.C.

Egel, K., August 1984. "Salvaging the Integrity of the RCS Program: The Santa Monica Energy Fitness Program," *Doing Better: Setting an Agenda for the Second Decade*, American Council for an Energy-Efficient Economy, Washington, D.C.

Energy Conservation Digest November 5, 1984. "Home Energy Rating System Proves Elusive, but Efforts and Pressures Continue at NIBS," *Energy Conservation Digest* VII(21).

Energy Information Administration September 1985. *An Economic Evaluation of Energy Conservation and Renewable Energy Tax Credits*, U.S. Department of Energy, Washington, D.C.

Esposito, B. August 1985. Minneapolis Energy Office, Minneapolis, Minn., personal communication to Howard Geller.

Florida Department of Community Affairs September 1982. *Model Energy Efficiency Code for Building Construction,* Florida Department of Community Affairs, Tallahassee, Fla.

Ferrey, S. August 1984. "Pulling a Rabbit Out of the Hat: Innovative Financing for Low-Income Conservation," *Doing Better: Setting an Agenda for the Second Decade,* American Council for an Energy-Efficient Economy, Washington, D.C.

Frahm, A. August 1984. "Developing Energy Efficiency Guidelines for Residential Underwriting: the Shelter Industry Program," *Doing Better: Setting an Agenda for the Second Decade,* American Council for an Energy-Efficient Economy, Washington, D.C.

Gee, P. August 1984. "Energy Efficiency in Local Government Operations: the Local Government Energy Officer Program," *Doing Better: Setting an Agenda for the Second Decade,* American Council for an Energy-Efficient Economy, Washington, D.C.

Geller, H. S. June 1983. *Energy Efficient Appliances,* American Council for an Energy-Efficient Economy and the Energy Conservation Coalition, Washington, D.C.

Geller, H., and Miller, A. November–December 1984. "Efficient Appliances: Performance Issues and Policy Options," *Energy Conservation Bulletin* 4(3), 1–5.

General Accounting Office October 26, 1981. *Uncertain Quality, Energy Savings, and Future Production Hamper the Weatherization Program,* U.S. General Accounting Office, Washington, D.C.

General Accounting Office March 11, 1982a. *Studies on Effectiveness of Energy Tax Incentives Are Inconclusive,* U.S. General Accounting Office, Washington, D.C.

General Accounting Office May 1982b. *Appliance Efficiency Standards: Issues Needing Resolution by DOE,* GAO/EMD-82-78, U.S. General Accounting Office, Washington, D.C.

General Accounting Office April 1982c. *State Energy Conservation Program Needs Reassessing,* GAO/EMD-83-39, U.S. General Accounting Office, Washington, D.C.

Goldman, C. A. August 1984. "Measured Energy Savings from Residential Retrofits: Updated Results from the BECA-B Project," *Doing Better: Setting an Agenda for the Second Decade,* American Council for an Energy-Efficient Economy, Washington, D.C.

Goldstein, D. B. February/March 1983. "Refrigerator Reform: Guidelines for Energy Gluttons," *Technology Review* 86(2), 36–46.

Griffin, T., et al. August 1984. "An Energy Management Service for Multi-Family Rental Property in St. Paul, Minnesota," *Doing Better: Setting an Agenda for the Second Decade*, American Council for an Energy-Efficient Economy, Washington, D.C.

Haun, C. R. August 1984. "Seven Keys to Energy Conservation in Multi-Family Buildings: 'Citizens Conservation Corporation's Approach to Energy Conservation in Multi-Family Buildings Housing Low-Income and Elderly Residents,'" *Doing Better: Setting an Agenda for the Second Decade*, American Council for an Energy-Efficient Economy, Washington, D.C.

Hewitt, D., et al. October 31, 1984. *Low-income Weatherization Program Study*, Wisconsin Energy Conservation Corporation, Madison, Wis.

Hirst, E., Goeltz, R., and Manning, H. 1983. "Analysis of Household Retrofit Expenditures," *Energy Systems and Policy* 7, 303–322.

Hirst, E., August 1984. "Evaluation of Utility Home Energy Audit (RCS) Programs," *Doing Better: Setting an Agenda for the Second Decade*, American Council for an Energy-Efficient Economy, Washington, D.C.

Hirst, E. 1985. "Improving Energy Efficiency of Existing Homes: The Residential Conservation Service," in *State Energy Policy: Current Issues, Future Directions*, Westview Press, Boulder, Colo.

Hirst, E., and Armstrong, J. November 14, 1980. "Managing State Energy Conservation Programs," *Science* 210, 740-744.

Housing and Urban Development December 1984. *Solar Energy and Energy Conservation Bank Annual Report to Congress*, U.S. Dept. of Housing and Urban Development, Washington, D.C.

Kelly, H. 1984. Testimony before the Subcommittee on Housing and Community Development, Committee on Banking, Finance and Urban Affairs, U.S. House of Representatives, March 21–22.

Krumholtz, N., and McDermott, M. August 1984. "Poor People, Neighborhood Groups and Energy Conservation: a Cleveland Study," *Doing Better: Setting an Agenda for the Second Decade*, American Council for an Energy-Efficient Economy, Washington, D.C.

Kushler, M., Witte, P., and Crandall, G. C. 1984. "RCS in Michigan: Positive Results to Date—Where Do We Go from Here?" *Doing Better: Setting an Agenda for the Second Decade*, American Council for an Energy-Efficient Economy, Washington, D.C.

League of California Cities March 1984. *Local Energy Strategies: A Second Look*, League of California Cities, Sacramento, Calif.

Levine, M. D., et al. August 1984. "Analysis of Federal Appliance Energy Efficiency Standards," *Doing Better: Setting an Agenda for the Second Decade*, American Council for an Energy-Efficient Economy, Washington, D.C.

McCarty, K. S., and Willner, A. January 1985. *Home Energy Rating Systems: Purpose, Operations, Barriers and Future Research Needs,* Consumer Energy Council of America Research Foundation, Washington, D.C.

McCold, L. N., et al. February 1984. *Thermal Efficiency Standards and Codes, Volume 2: Relationships of ASHRAE Standards and External Factors to Energy Efficient Practices in New Homes*, ORNL/CON-1-1/V2, Oak Ridge National Laboratory, Oak Ridge, Tenn.

McNeill, D. H., and Wilkie, W. L. 1979. "Public Policy and Consumer Information: Impact of the New Energy Labels," *Journal of Consumer Research* 6(1), 1–11.

Mihaly, M. E. August 1984. "No Cost/Low Cost: a Community Energy-Saving Effort," *Doing Better: Setting an Agenda for the Second Decade*, American Council for an Energy-Efficient Economy, Washington, D.C.

Nisson, J. D. December 1984. "New South Dakota Law Mandates Superinsulation, Blower Door Tests, and Mechanical Ventilation," *Energy Design Update* 1(1), 27–31.

Northwest Power Planning Council April 1983. *Northwest Conservation and Electric Power Plan,* Northwest Power Planning Council, Portland, Oreg.

Peabody, G. E. August 20, 1984. *Weatherization Program Evaluation,* SR-EEUD-84-1, Energy Information Administration, U.S. Department of Energy, Washington, D.C.

Preysner, W. R. May 1985. U.S. Department of Housing and Urban Development, Washington, D.C., personal communication to Howard Geller.

Proctor, J. August 1984. "Low Cost Furnace Efficiency Improvements," *Doing Better: Setting an Agenda for the Second Decade*, American Council for an Energy-Efficient Economy, Washington, D.C.

Randolph, J. December 1984. "Energy Conservation Programmes—A Review of State Initiatives in the USA," *Energy Policy* 12(4), 425–438.

Ritschard, R. L. et al. August 1984. "Energy Calculation Slide Rules for Single Family Houses," *Doing Better: Setting an Agenda for the Second Decade*, American Council for an Energy-Efficient Economy, Washington, D.C.

Rollin, P., and Beyea, J. October 1985, "U.S. Appliance Efficiency Standards," *Energy Policy* 13(5), 425–536.

Rosenfeld, A. H., and Schuck, L. March–April 1984. "Home Energy Ratings: A Promising Approach to Energy Conservation," *Energy Conservation Bulletin* 3, 5.

Sawyer, S. W., August 1984. "Improving Federal-State Energy Conservation Programs: the States' Perspective," *Doing Better: Setting an Agenda for the Second Decade*, American Council for an Energy-Efficient Economy, Washington, D.C.

Sawyer, S. W. April 1985. "Federal-State Conservation Programs," *Energy Policy* 13(2), 156–168.

Sawyer, S. W., and Armstrong, J. R. 1985. *State Energy Policy: Current Issues, Future Directions,* Westview Press, Boulder, Colo.

Stern, P. C., and Aronson, E., eds. 1984. *Energy Use: The Human Dimension,* W. H. Freeman and Co., New York.

U.S. Court of Appeals July 16, 1985. *Natural Resources Defense Council, Inc., vs U.S. Department of Energy,* 83-1195, District of Columbia Circuit Court, Washington, D.C.

Veigel, J. W. August 1984. "A Voluntary State/Utility Partnership: the North Carolina Alternative Energy Corporation," *Doing Better: Setting an Agenda for the Second Decade*, American Council for an Energy-Efficient Economy, Washington, D.C.

Veigel, J., and Lakoff, S. October 1985. "U.S. States and Energy Efficiency: North Carolina's 'Quango'," *Energy Policy* 13(5), 445–457.

Walker, J. A., Rauh, T. N., and Griffin, K. 1985. "A Review of the Residential Conservation Service Program," *Annual Review of Energy, 10,* Annual Reviews, Palo Alto, Calif.

Weedall, M. August 1984. "The Emerging Role of the Public Sector in Third-Party Finance," *Doing Better: Setting an Agenda for the Second Decade*, American Council for an Energy-Efficient Economy, Washington, D.C.

Western Sun 1979. *Capturing the Sun's Energy—Opportunities for Local Governments,* Western Sun, Sacramento, Calif.

Weisenmiller, R. B. 1984. "Status of Third-Party Financing for Efficiency Improvements in Buildings," *Doing Better: Setting an Agenda for the Second Decade*, American Council for an Energy-Efficient Economy, Washington, D.C.

Wicks, A. August 1984. "Consistency and Support: Key to Effective Implementation of Conservation Regulations," *Doing Better: Setting an Agenda for the Second Decade*, American Council for an Energy-Efficient Economy, Washington, D.C.

Woodham, B. G. March 21–22, 1984. Testimony before the Subcommittee on Housing and Community Development, Committee of Banking, Finance and Urban Affairs, U.S. House of Representatives, Washington, D.C.

CHAPTER 10

American Gas Association April 12, 1984. "Gas Consumption by Residential Appliances," *Public Utilities Fortnightly* **113**, 8.

Argonne National Laboratory August 1985. *Energy Conservation Program Evaluation: Practical Methods, Useful Results, Proceedings of the Second National Conference,* Argonne National Laboratory, Argonne, Ill.

Berry, L., et al. August 1983. *Design Options to Test the Effects of Financial Incentives in a Utility Conservation Program: TVA's Heat Pump Water Heater Program,* ORNL/CON-125, Oak Ridge National Laboratory, Oak Ridge, Tenn.

Brown, G., and Levett, J. April 12, 1984. "Conservation in Perspective," *Public Utilities Fortnightly* **113**, 8.

Brown, M. A., and Reeves, G. July 1985. *The Implementation Phase of a Residential Energy Conservation Shared Savings Program: The General Public Utilities Experience,* ORNL/CON-187, Oak Ridge National Laboratory, Oak Ridge, Tenn.

Caldwell, R. 1984. Remarks made during the EPRI Demand-Side Management: Market Acceptance, Load Shape Impacts, and Planning Techniques seminar, New Orleans, September 26–27.

Calhoun, R. August 1984. "The Great PG and E Energy Rebate," *Doing Better: Setting an Agenda for the Second Decade,* American Council for an Energy-Efficient Economy, Washington, D.C.

California Public Utilities Commission October 1984. *1983 Energy Conservation Program Summary,* San Francisco, Calif.

Caves, D., and Christensen, L. March 17, 1983. "Time-of-use Rates for Residential Electric Service: Results from The Wisconsin Experiment," *Public Utilities Fortnightly* **111**, 6.

Collins, N. E., et al. April 1985. *Past Efforts and Future Directions for Evaluating State Energy Conservation Programs,* ORNL-6113, Oak Ridge National Laboratory, Oak Ridge, Tenn.

Davis, T., et al. August 1984. "Utility Experience in Designing, Implementing and Marketing Demand-Side Programs," *Doing Better: Setting an Agenda for the Second Decade,* American Council for an Energy-Efficient Economy, Washington, D.C.

Department of Energy January 1984a. *Residential Conservation Service Evaluation Report,* Division of Building Services, U.S. Department of Energy, Washington, D.C.

Department of Energy November 1984b. *Fourth Annual Report on Financing, Supply, and Installation Activities of Public Utilities in Connection with the Residential Conservation Service Program,* U.S. Department of Energy, Washington, D.C.

Dickey, D. F., et al. August 1984. "Effects of Utility Incentive Programs for Appliances on the Energy Efficiency of Newly Purchased Appliances," *Doing Better: Setting an Agenda for the Second Decade,* American Council for an Energy-Efficient Economy, Washington, D.C.

Egel, K. August 1984. "Salvaging the Integrity of the RCS Program: The Santa Monica Energy Fitness Program," *Doing Better: Setting an Agenda for the Second Decade,* American Council for an Energy-Efficient Economy, Washington, D.C.

Electric Power Research Institute May 1984a. *Utility Conservation Programs: Planning, Analysis and Implementation,* EPRI EA-3530, Electric Power Research Institute, Palo Alto, Calif.

Electric Power Research Institute July 1984b. *Electric Utility Conservation Program: Assessment of Implementation Experience. Volume 1: Project Overview and Major Findings; Volume 2: Project Results,* EPRI EA-3585, Electric Power Research Institute, Palo Alto, Calif.

Electric Power Research Institute 1984c. Conference Notebook for *Demand-Side Management: Market Acceptance, Load Shape Impacts, and Planning Techniques,* Electric Power Research Institute, Palo Alto, Calif.

Elliott, H. N. February 1985. Marketing Planning Coordinator, Florida Power and Light Co., Miami, personal communication to Howard Geller.

Esteves, R. September 1983. "Don't Pay for Insulation ...Buy Conservation," presented at 1983 EPRI Conference on Utility Conservation Programs, New Orleans.

Faruqui, A., Aigner, D., and Howard, R. March 1981. *Customer Response to Time-of-Use Rates,* RDS-84, Electric Power Research Institute, Palo Alto, Calif.

Faruqui, A., and Malko, J. R. 1983. "The Residential Demand for Electricity by Time-of-Use: A Survey of Twelve Experiments with Peak Load Pricing," *Energy* **8**, 10.

Fels, M., and Kempton, W. May 1984. *Toward More Informative Energy Bills,* PU/CEES Report No. 164, Michigan Agricultural Experiment Station, East Lansing, Mich.

Flaherty, W. December 1984. Public Staff, California Public Utilities Commission, personal communication to Jeanne Clinton.

Geller, H. June 1983. *Review of Utility Appliance Rebate Programs and Assessment of Their Potential in the Pacific Northwest,* American Council for an Energy-Efficient Economy, Washington, D.C.

Glazer, S. July-August 1984. "The Residential Conservation Service: Expectations, Performance, and Potential for the Future," *Energy Conservation Bulletin* **4**, 1.

Gunn, R. H. October 1983. *Northern States Power Company's Evaluation of the Appliance Rebate Program,* Northern States Power, Minneapolis, Minn.

Hirst, E. September/October 1984. "Household Energy Conservation: A Review of the Federal Residential Conservation Service Program," *Public Administration Review* **44**(5), 421–430.

Investor Responsibility Research Center September 1983. *Generating Energy Alternatives: Conservation, Load Management, and Renewable Energy at America's Electric Utilities,* Investor Responsibility Research Center, Washington, D.C.

Kuliasha, M. A., et al. May 1985. *Field Performance of Residential Thermal Storage Systems,* EM-4041, Electric Power Research Institute, Palo Alto, Calif.

Kushler, M. August 1984. "RCS in Michigan: Positive Results to Date—Where do We go from Here?," *Doing Better: Setting an Agenda for the Second Decade*, American Council for an Energy-Efficient Economy, Washington, D.C.

Lehenbauer, G. September 13–15 1983. "Utility Financing of Energy Management: The Puget Sound Power and Light Experience," presented at the EPRI, Alliance to Save Energy, and Synergic Resources Corporation Conference on Utility Conservation Programs: Planning, Analysis and Implementation, New Orleans.

Lucchi, S. December 1984. Supervising Utilities Engineer, California Public Utilities Commission, personal communication to Jeanne Clinton.

McDonough, H., and Parisi, D. H. August 1984. "Communities, Contractors, and Residents—3,000 Attics Later," *Doing Better: Setting an Agenda for the Second Decade*, American Council for an Energy-Efficient Economy, Washington, D.C.

Morgenstern, R., and Dubinsky, R. January 20, 1983. "A Utility-Financed Weatherization Program in the Mid-Atlantic Region: The Economics," *Public Utilities Fortnightly* **111**, 2.

Moulton, D. August 1984. "The Impact of Utility-Sponsored Energy Conservation Loan Programs on Low-Income Households," *Doing Better: Setting an Agenda for the Second Decade*, American Council for an Energy-Efficient Economy, Washington, D.C.

Naill, R., and Sant, R. April 26, 1984. "Electricity Markets in the 1990's: Feast or Famine?" *Public Utilities Fortnightly* **113**, 9.

O'Keefe, J. September 13–15, 1983. "Measuring the Impact of Residential Conservation Programs: A Case Study," presented at the EPRI, Alliance to Save Energy, and Synergic Resources Corporation Conference on Utility Conservation Programs: Planning, Analysis and Implementation, New Orleans.

Sanghvi, A. March 1984. "Least-Cost Energy Strategies for Power System Expansion," *Energy Policy* **12**, 1.

Sant, R. 1979. *The Least-Cost Energy Strategy,* The Energy Productivity Center, Arlington, Va.

Savitz, M., et al. January 1984. *Utility Appliance Rebate Programs,* North Carolina Alternative Energy Corporation, Research Triangle Park, N.C.

Sawhill, J., and Silverman, L. May 26, 1983. "Build Flexibility—Not Power Plants," *Public Utilities Fortnightly* **111**(11), 17–21.

Schick, S. August 1984. "Learning About Commercial Program Design—BPA Purchase of Energy Savings," *Doing Better: Setting an Agenda for the Second Decade*, American Council for an Energy-Efficient Economy, Washington, D.C.

Schrader, T. August 1984. *Shared Savings Program, An Energy Conservation Program of the Wisconsin Gas Company, Report on the Pilot Project,* Wisconsin Gas Company, Milwaukee, Wis.

Stern, P., and Aronson, E. 1984. *Energy Use: The Human Dimension,* W. H. Freeman and Co., New York.

Tennessee Valley Authority May 1984. *Energy Conservation Program,* Division of Conservation and Energy Management, Knoxville, Tenn.

Tonn, B. E., Holub, E., and Hilliard, M. January 1986. *The Bonneville Power Administration Conservation/Load/ Resource Modeling Process: Review, Assessment, and Suggestions for Improvement,* ORNL/CON-190, Oak Ridge National Laboratory, Oak Ridge, Tenn.

U.S. Congress, House Committee on Small Business, May 1984. *Competition Between Small Business and Public Utilities,* Subcommittee on Antitrust and Restraint of Trade Activities Affecting Small Business, U.S. House of Representatives, USGPO, Washington, D.C.

Williams, M. September 13–15, 1983. "Segmenting Markets for Conservation," presented at the EPRI, Alliance to Save Energy, and Synergic Resources Corporation Conference on Utility Conservation Programs: Planning, Analysis and Implementation, New Orleans.

Wirtshafter, R. January 1983. "Energy Conservation in Gas Utilities: An Opportunity for Industry," *Public Utilities Fortnightly* 111, 2.

CHAPTER 11

Anachem, Inc., and Sandia National Laboratories September 1982. *Indoor Air Quality Handbook,* SAND82-17773, Sandia National Laboratories, Albuquerque, N.M.

American Society of Heating, Refrigerating, and Air-Conditioning Engineers 1981. *Ventilation for Acceptable Indoor Air Quality, Standard 621981,* American Society of Heating, Refrigerating, and Air-Conditioning Engineers, Atlanta, Ga.

Bernstein, R. S., Folk, H., Turner, D. R., and Melius, J. September 1984. "Nonoccupational Exposures to Indoor Air Pollutants: A Survey of State Programs and Practices," *American Journal of Public Health* 74(9), 1020–1023.

Dumont, R. 1983. *Building Research Note 212,* National Research Council of Canada, Ottawa, Ont.

Everett, J. J., and Dreher, T. J. 1982. "Institutional Aspects of Indoor Air Pollution in Energy Efficient Residences," *Environment International* 8, 525–531.

Federal Register August 9, 1984. "Manufactured Home Construction and Safety Standards, Final Rule," *Federal Register,* 31996.

Fisk, W. J., and Turiel, Y. February 1982. *Residential Air-to-Air Heat Exchangers: Performance, Energy Savings, and Economics,* LBL-13843, Lawrence Berkeley Laboratory, Berkeley, Calif.

Housing and Urban Development April 1981. *Mobile Home Research—An Evaluation of Formaldehyde Problems in Residential Mobile Homes,* U.S. Department of Housing and Urban Development, Washington, D.C.

Int-Hout, D. August 1984. "Environmental Control with Variable Air Volume Systems," *Doing Better: Setting an Agenda for the Second Decade*, American Council for an Energy-Efficient Economy, Washington, D.C.

Janssen, J. E. August 1984. "The ASHRAE Ventilation Standard 62-1981: A Status Report," presented at the 3rd International Conference on Indoor Air Quality and Climate, Stockholm.

Lawrence Berkeley Laboratory February 1984. *Manual on Indoor Air Quality*, EPRI EM-3469, Electric Power Research Institute, Palo Alto, Calif.

Macriss, R. A., and Elkens, R. H. May 1977. "Control of the Level of NO$_x$ in the Indoor Environment," presented at the Fourth International Clean Air Congress, Tokyo.

Macriss, R. A. October 1983. "Air Infiltration and Indoor Air Quality," *Proceedings of an Engineering Foundation Conference on Management of Atmospheres in Tightly Enclosed Spaces*, American Society of Heating, Refrigerating and Air-Conditioning Engineers, Atlanta, Ga.

Meier, A., ed. January-February 1985. "The Kerosene Heater Controversy," *Energy Auditor and Retrofitter* 1(1), 27–31.

Meyer, B. August 1984. "Formaldehyde Release from Building Products," presented at the 3rd International Conference on Indoor Air Quality and Climate, Stockholm.

Meyer, B., and Hermanns, K. June 1984. "Formaldehyde Indoor Air Problems," presented at the 77th Annual Meeting of the Air Pollution Control Association, San Francisco.

Morrill, J., and Geller, H. April 1985. *Residential Indoor Air Quality in North Carolina,* North Carolina Alternative Energy Corporation, Research Triangle Park, N.C.

National Academy of Sciences 1981. *Indoor Pollutants,* Committee on Indoor Pollutants, National Research Council, National Academy Press, Washington, D.C.

National Center for Appropriate Technology March 1984. *Heat Recovery Ventilation for Housing,* DOE/CE/15095-9, U.S. Department of Energy, Washington, D.C.

National Council on Radiation Protection May 1984. *Evaluation of Occupational and Environmental Exposures to Radon and Radon Daughters in the United States*, National Council on Radiation Protection and Measurements, Bethesda, Md.

Nero, A. V., et al. August 1983. "Radon Concentrations and Infiltration Rates Measured in Conventional and Energy Efficient Houses," *Health Physics* **45**, 401–405.

Nisson, J. D., ed. December 1984. "New South Dakota Law Mandates Superinsulation, Blower Door Tests, and Mechanical Ventilation," *Energy Design Update* **1**(1), 27–31.

Nisson, J. D., and Dutt, G. S. February 1985. *The Super-insulated Home Book*, John Wiley and Sons, New York.

Nitschke, I. A., et al. August 1984. "A Detailed Study of Inexpensive Radon Control Techniques in New York State Houses," presented at the 3rd International Conference on Indoor Air Quality and Climate, Stockholm.

Ontario Research Foundation 1984. *Laboratory Evaluation of Air-to-Air Heat Exchanger Performance*, Ontario Research Foundation, Mississauga, Ont., Canada.

Repace, J. L. 1982. "Indoor Air Pollution," *Environment International* **8**, 21–36.

Riley, M. August 1984. Dept. of Energy, Mines and Resources, Ottawa, Canada, personal communication to Howard Geller.

Sachs, H. M., and Hernandez, T. L. June 1984. "Residential Radon Control by Subslab Ventilation," presented at the 77th Annual Meeting of the Air Pollution Control Association, San Francisco, Calif.

Sawyer, R. N., and Spooner, C. M. March 1978. *Sprayed Asbestos-Containing Materials: A Guidance Document,* EPA-450/2-78-014, U.S. Environmental Protection Agency, Washington, D.C.

Scott, A. G., and Findlay, W. O. July 1983. *Demonstration of Remedial Techniques Against Radon in Houses on Florida Phosphate Lands,* EPA-520/5-83/009, U.S. Environmental Protection Agency, Washington, D.C.

Sexton, K., and Wesolowski, J. J. 1985. "Safeguarding Indoor Air Quality: Public Policy Issues and the California Response," *Environmental Science and Technology* **19**(4), 305–309.

Sextro, R. G., et al. August 1984. "Evaluation of Indoor Aerosol Control Devices and their Effects on Radon Progeny Concentrations," presented at the 3rd International Conference on Indoor Air Quality and Climate, Stockholm.

Shukla, K. C., and Hurley, J. R. July 1983. *Development of an Efficient, Low NO_x Domestic Gas Range Cook Top*, TE4311-143-83, Thermo Electron Corp., Waltham, Mass.

Spengler, J. D., and Sexton, K. 1983. "Indoor Air Pollution: A Public Health Perspective," *Science* **221**, 9–17.

Stolwijk, J. August 1984. "The Sick Building Syndrome," presented at the 3rd International Conference on Indoor Air Quality and Climate, Stockholm.

Thor, P. W. August 1984. "BPA Radon Field Monitoring Study," *Doing Better: Setting an Agenda for the Second Decade*, American Council for an Energy-Efficient Economy, Washington, D.C.

Traynor, G. W., et al. 1982. "The Effects of Ventilation on Residential Air Pollution Due to Emmissions from a Gas-Fired Range," *Environment International* **8**, 447–452.

Treitman, R. D., and Spengler, J. D. August 1984. "Equipment for Personal and Portable Air Monitoring: A State-of-the-Art Survey and Review," presented at the 3rd International Conference on Indoor Air Quality and Climate, Stockholm.

Ulsamer, A. G., et al. August 1984. "Health Effects of Formaldehyde: An Indoor Pollutant," presented at the 3rd International Conference on Indoor Air Quality and Climate, Stockholm.

CHAPTER 12

American Institute of Architects 1984. *Architectural Research Priorities,* American Institute of Architects Foundation, Washington, D.C.

Committee on Science and Technology May 1983. *Building Energy Research Workshop,* U.S. House of Representatives, USGPO, Washington, D.C.

Committee on Science and Technology April 1984. *Building Equipment Research Workshop,* U.S. House of Representatives, USGPO, Washington, D.C.

Department of Energy July 1984. *Energy Conservation Multi-Year Plan, FY1986—FY1990,* U.S. Department of Energy, Washington, D.C.

Electric Power Research Institute June 1984. *Residential and Commercial Programs, Program Overview and Project Descriptions,* Electric Power Research Institute, Palo Alto, Calif.

Electric Power Research Institute December 1985. *Electricity Outlook: The Foundation of EPRI R&D Planning,* Electric Power Research Institute, Palo Alto, Calif.

Gas Research Institute September 1984. *Residential/Commercial Energy Systems Research Status Report, 1983-1984,* Gas Research Institute, Chicago, Ill.

Oak Ridge National Laboratory March 1982. *The National Program Plan for the Thermal Performance of Building Envelope Systems and Materials,* ORNL/Sub-7973/1, Oak Ridge National Laboratory, Oak Ridge, Tenn.

CHAPTER 13

American Institute of Architects Foundation 1983. *Technology Transfer from the National Laboratories to the Building Industry,* American Institute of Architects, Washington, D.C.

Arbor, Inc. 1984. *1983 Southern California Edison Customer Attitude Survey,* Southern California Edison Company, Rosemead, Calif.

Argonne National Laboratory August 1985. *Proceedings of Conference on Energy Conservation Program Evaluation: Practical Methods, Useful Results,* Argonne National Laboratory, Argonne, Ill.

American Society of Heating, Refrigerating, and Air-Conditioning Engineers January 22, 1985. *Proceedings of Building Industry Roundtable on Technology Transfer and Research Utilization,* Battelle Pacific Northwest Laboratory, Richland, Wash.

Booz Allen & Hamilton, Inc. February 28, 1985. *Customer Preference and Behavior,* proposal submitted to Electric Power Research Institute, Palo Alto, Calif.

Brown, M., et al. 1985. *A Strategy for Accelerating the Use of Energy Conserving Building Technologies,* ORNL/TM-9630, Oak Ridge National Laboratory, Oak Ridge, Tenn.

Clinton, J. 1985. "Energy Management in the Commercial Sector: A Survey of Professional Energy Managers," unpublished manuscript.

Morrison, B., and Kempton, W. 1984. *Families and Energy: Coping with Uncertainty,* conference proceedings, Institute for Family and Child Study, Michigan State University, East Lansing, Mich.

Pacific Gas and Electric Company May 1985. *Meeting Energy Challenges, Proceedings of the Great PG&E Energy Expo,* Pergamon Press, New York.

Science Applications, Inc. 1983. *Proceedings of the Building Energy Research Workshop,* Committee on Science and Technology, U.S. House of Representatives, USGPO, Washington, D.C.

Sorenson, J. April 1985. *Information and Energy Conserving Behavior: Past Efforts and Future Directions for Evaluating State Energy Conservation Programs,* ORNL-6113, Oak Ridge National Laboratory, Oak Ridge, Tenn.

Stern, P., and Aronson, E. 1984. *Energy Use: The Human Dimension,* W. H. Freeman and Co., New York.

Synergic Resources Corporation 1985. *The Feasibility of Developing an Information Clearinghouse on Energy Efficiency Programs,* draft, U.S. Department of Energy, Washington, D.C.

Synergic Resources Corporation May 1984. *Conference on Utility Conservation Programs: Planning, Analysis, and Implementation,* EPRI EA-3530, Electric Power Research Institute, Palo Alto, Calif.

Williams, M. 1983. "Segmenting Markets for Conservation," presented at the EPRI, Alliance to Save Energy, Synergic Resources Corporation Conference Utility Conservation Programs: Planning, Analysis and Implementation, New Orleans.

CHAPTER 14

Stern, P., and Aronson, E. 1984. *Energy Use: The Human Dimension,* W. H. Freeman and Co., New York.

INDEX